INSECTS AND OTHER TERRESTRIAL ARTHROPODS: BIOLOGY, CHEMISTRY AND BEHAVIOR

BEE HEALTH

FACTORS, ANALYSES, AND RESEARCH PROGRESS

INSECTS AND OTHER TERRESTRIAL ARTHROPODS: BIOLOGY, CHEMISTRY AND BEHAVIOR

Additional books in this series can be found on Nova's website under the Series tab.

Additional E-books in this series can be found on Nova's website under the E-book tab.

GLOBAL AGRICULTURE: DEVELOPMENTS, ISSUES, AND RESEARCH

Additional books in this series can be found on Nova's website under the Series tab.

Additional E-books in this series can be found on Nova's website under the E-book tab.

INSECTS AND OTHER TERRESTRIAL ARTHROPODS: BIOLOGY, CHEMISTRY AND BEHAVIOR

BEE HEALTH

FACTORS, ANALYSES, AND RESEARCH PROGRESS

ROGER DUNHAM
EDITOR

nova publishers
New York

For permission to use material from this book please contact us:
Telephone 631-231-7269; Fax 631-231-8175
Web Site: http://www.novapublishers.com

Additional color graphics may be available in the e-book version of this book.

Library of Congress Cataloging-in-Publication Data

ISBN: 978-1-62618-044-4

Published by Nova Science Publishers, Inc. † New York

CONTENTS

PREFACE

Bees, both commercially managed honey bees and wild bees, play an important role in global food production. In the United States, the value of honey bees only as commercial pollinators in U.S. food production is estimated at about $15 billion to $20 billion annually. Given their importance to food production, many have expressed concern about whether a "pollinator crisis" has been occurring in recent decades. Worldwide reports indicate that populations of both managed honey bees and native bees have been declining, with colony losses in some cases described as severe or unusual. Aside from the range of potential identified causes for bee colony declines, this book addresses what role, if any, pesticides play in influencing the health and wellness of bees. The book begins by describing what is known about managed and wild bee populations, continues with brief descriptions of factors that might be affecting those populations, including a more detailed look at the potential effects of pesticide exposures and risk management.

Chapter 1 - Bees, both commercially managed honey bees and wild bees, play an important role in global food production. In the United States, the value of honey bees only as commercial pollinators in U.S. food production is estimated at about $15 billion to $20 billion annually. The estimated value of other types of insect pollinators, including wild bees, to U.S. food production is not available. Given their importance to food production, many have expressed concern about whether a "pollinator crisis" has been occurring in recent decades. In the United States, commercial migratory beekeepers along the East Coast of the United States began reporting sharp declines in 2006 in their honey bee colonies. The U.S. Department of Agriculture (USDA) reports that overwinter colony losses from 2006 to 2011 averaged more than 32% annually. This issue remained legislatively active in the 110th Congress and resulted in increased funding for pollinator research, among other types of farm program support, as part of the 2008 farm bill (P.L. 110-246). Congressional interest in the health of honey bees and other pollinators has continued in the 112th Congress (e.g., H.R. 2381, H.R. 6083, and S. 3240) and may extend into the 113th Congress.

This report:

- Describes changes in managed and wild bee populations, given readily available data and information. It focuses on managed and wild bees only, and excludes other types of pollinators, including other insects, birds, and bats. Data on managed honey bees are limited, and do not provide a comprehensive view of changes in bee populations. Data for wild bee populations are even more limited.

- Provides a listing of the range of possible factors thought to be negatively affecting managed and wild bee populations. In addition to pesticides, other identified factors include bee pests and diseases, diet and nutrition, genetics, habitat loss and other environmental stressors, and beekeeping management issues, as well as the possibility that bees are being negatively affected by cumulative, multiple exposures and/or the interactive effects of each of these factors.
- Briefly summarizes readily available scientific research and analysis regarding the potential role of pesticides among the factors affecting the health and wellbeing of bees, as well as the statutory authority and related regulatory activities of the U.S. Environmental Protection Agency (EPA) related to pesticide use.

A 2007 report by the National Research Council of the National Academy of Sciences, *Status of Pollinators in North America,* provides a more detailed scientific context for this report and may be consulted for more in depth understanding about bee health. That study concluded that many factors contribute to pollinator declines in North America, and CRS accedes to that conclusion.

Accordingly, the focus of this report on bee exposure to pesticides is not intended to imply that pesticides are any more important in influencing the health and wellness of bees than any of the other identified factors influencing bee health. Pesticides are only one of the many influences on bee health.

Because neonicotinoid pesticides have been the focus of concerns in Europe and in the United States, this report briefly describes recent scientific research related to possible effects of exposure to these pesticides on bees.

The report concludes with a summary of recent regulatory activity regarding neonicotinoids at EPA, the federal agency charged with assessing risks and regulating U.S. sale and use of pesticides.

Chapter 2 - Starting in late 2006, commercial migratory beekeepers along the East Coast of the United States began reporting sharp declines in their honey bee colonies. Because of the severity and unusual circumstances of these colony declines, scientists named this phenomenon colony collapse disorder (CCD). Reports indicate that beekeepers in most states have been affected. Overall, the number of managed honey bee colonies dropped an estimated 35.8% and 31.8% in the winters of 2007/2008 and 2006/2007, respectively. Preliminary loss estimates for the 2008/2009 winter are reported at 28.6%. To date, the precise reasons for colony losses are not yet known.

Honey bees are the most economically valuable pollinators of agricultural crops worldwide. Scientists at universities and the U.S. Department of Agriculture (USDA) frequently assert that bee pollination is involved in about one-third of the U.S. diet, and contributes to the production of a wide range of fruits, vegetables, tree nuts, forage crops, some field crops, and other specialty crops. The monetary value of honey bees as commercial pollinators in the United States is estimated at about $15-$20 billion annually.

Honey bee colony losses are not uncommon. However, losses in recent years differ from past situations in that colony losses are occurring mostly because bees are failing to return to the hive (which is largely uncharacteristic of bee behavior); bee colony losses have been rapid; colony losses are occurring in large numbers; and the reason(s) for these losses remains largely unknown.

Based on the available research over the past few years on the numerous possible causes of CCD, USDA concluded in its 2007-2008 progress report (released in June 2009) that "it now seems clear that no single factor alone is responsible for the malady." This has led researchers to further examine the hypothesis that CCD may be "a syndrome caused by many different factors, working in combination or synergistically." Currently, USDA states, researchers are focusing on three major possibilities:

- pesticides that may be having unexpected negative effects on honey bees;
- a new parasite or pathogen that may be attacking honey bees, such as the parasite *Nosema ceranae* or viruses; and
- a combination of existing stresses that may compromise the immune system of bees and disrupt their social system, making colonies more susceptible to disease and collapse. Stresses could include high levels of infection by the *Varroa* mite; poor nutrition due to apiary overcrowding, pollination of crops with low nutritional value, or pollen or nectar scarcity; exposure to limited or contaminated water supplies; and migratory stress.

Funding for honey bee and CCD research at USDA's Agricultural Research Service (ARS) has increased sharply, following both the enactment of the 2008 farm bill (P.L. 110-246) and the FY2009 and FY2010 appropriations process (P.L. 111-8 and P.L. 111-80, respectively). These legislative actions contained additional provisions that would, among other things, provide additional funding for research and conservation programs addressing honey bees and pollinators. Total ARS funding for honey bee and CCD research averaged more than $7.7 million each in FY2007 and FY2008, increasing to $8.3 million in FY2009 and $9.8 million for FY2010.

Chapter 3 - When it comes to solving the puzzling syndrome known as "colony collapse disorder" (CCD), which has been attacking honey bee colonies since 2006, the best that can be said is that there is good news and bad news. The good news is that the rate of honey bee losses seems to have leveled off rather than continuing to increase. The bad news is that the cause or causes of CCD remain unclear.

In the United States, the problem surfaced in October 2006, when an increasing number of beekeepers began reporting losses of 30 to 90 percent of the hives in their apiaries with no apparent cause. The defining characteristic of CCD is the disappearance of most, if not all, of the adult honey bees in a colony, leaving behind honey and brood but no dead bee bodies. This definition has recently been revised to include low levels of Varroa mite and other pathogens, such as Nosema, as probable contributing factors.

For the last 3 years, self-reported surveys of beekeepers have found that winter colony losses have averaged about 30 percent, with around one-third of those losses ascribed to CCD, according to Jeff Pettis, research leader of the Bee Research Laboratory in Beltsville, Maryland, who heads up the Agricultural Research Service's CCD research effort. This compares to colony losses that were averaging 15-20 percent before CCD.

Chapter 4 - The 2011 USDA Animal Plant and Health Inspection (APHIS) sponsored National Survey of Honey Bee Pests and Diseases was conducted in collaboration with the University of Maryland (UMD) and the USDA Agricultural Research Service (ARS) and with the cooperation of 34 states covering the timeframe from the spring of 2011 through the summer of 2012. This survey is in its third year and is the largest to date. The National

Survey has expanded from a 3 state Pilot Survey in 2009 to a Limited National survey of 13 states in 2010. This expansion has allowed us to augment and extend the baseline pest and pathogen data collected from the previous 2 surveys. The primary focus of this survey was to verify the absence of exotic threats such as the parasitic mite, *Tropilaelaps* spp., *Apis cerana*, and Slow Paralysis Virus. Under current international trade agreements, the U.S. cannot deny import permits from other nations unless the exporting nation has a disease, parasite, or pest of honey bees that is not found in the U.S. Establishing the absence of threats to honey bee populations not thought to be present in the U.S. was the primary objective of this effort.

With exotic species as the primary target, a secondary objective was set to make use of the sampling by determining existing levels of other honey bee diseases and parasites known to be present in the U.S. This was also performed in the previous two surveys. The survey results are used to gauge the overall health of colonies, to create a baseline disease level, and to facilitate interpretation of ongoing and future epidemiological studies. This baseline data, including historic data from research institutions such as the ARS Bee Research Laboratory (BRL) and other ongoing field sampling and management surveys, have been incorporated into a single database as part of the Bee Informed Partnership (www.beeinformed.org), a 5 year grant funded by USDA NIFA (National Institute of Food and Agriculture). The 2011-2012 National Survey effort included collection of samples from 34 states: Alabama, Arkansas, California, Colorado, Delaware, Florida, Georgia, Hawaii, Idaho, Illinois, Indiana, Iowa, Louisiana, Maryland, Michigan, Minnesota, Montana, Nebraska, New York, New Mexico, New Hampshire, New Jersey, North Carolina, North Dakota, Ohio, Pennsylvania, South Carolina, South Dakota, Tennessee, Texas, Utah, Virginia, West Virginia and Wisconsin. A total of 875 samples representing 7,000 colonies are expected at the completion of this survey. To date, approximately 94% of the live bee samples (for virus analysis) have been collected and analyzed and about 93% of the alcohol samples (for Varroa mite, Nosema and Tropilaelaps determination) have been delivered and processed. Also at this time, 80% of the reports to beekeepers and apiary inspectors have been distributed. A second full National Survey for 2012 was initiated early this summer with 32 participating states. Samples for this fourth survey have already begun to arrive.

The survey samples were analyzed for 10 known honey bee viruses, pests and pathogens. Slow Bee Paralysis Virus (SBPV), the only virus included in this year's testing that is not currently found in the U.S., was examined in all samples and no detection was made. No diseases or parasites of bees not already known to exist in the country were discovered. Two common honey bee viruses, Deformed Wing Virus (DWV) and Black Queen Cell Virus (BQCV), were found in all 34 states. In this 2011 survey, 9 samples out of 673 tested positive for *Nosema apis*, or about 1.3% of the samples. This *Nosema* species was not detected in the previous survey years; however, this may be attributed to the development and implementation of a more sensitive primer used in the Polymerase Chain Reaction (PCR) detection this year. For the third year, we saw no evidence of *Tropilaelaps* mites. Since honey bee tracheal mites (*Acarapis woodi*) were not detected in samples in 2009 nor in 2010, samples were not analyzed for this mite this year. Honey bee tracheal mites are known to exist in the country and our failure to find them may be the result of our sampling procedure. Varroa mites continued to be observed in all states with the exception of the Hawaiian Islands of Maui, Kauai and Molokai.

New to the survey this year was a pilot pollen pesticide survey where a minimum of 3 grams of pollen was collected from brood frames that was tested for 174 known pesticides.

The pollen was collected from the same composite 8 colonies undergoing the standard survey sampling and sent to the USDA Agricultural Marketing Service (AMS) in Gastonia, NC for analysis. This pilot pesticide survey focused on 11 states: Alabama, California, Colorado, Florida, Idaho, Indiana, New York, South Dakota, Tennessee, Texas and Wisconsin. We received 99 samples from these states and of the 174 possible pesticides, only 35 were detected. This pollen survey will be expanded and will include each participating state in the 2012 National Survey.

This survey was designed to be representative of the managed honey bees across the broad geography of the United States. The survey was open to any state wishing to participate and beekeeper participation was completely voluntary and the beekeeper did not have to be present or assist with the sampling. The results can be considered as representative of the pests and pathogens present in the U.S.

Chapter 5 - This report describes research progress on Honey Bee Colony Collapse Disorder during 2011. Mandated by Congress in the 2008 Farm Bill (section 7204[h][4]), this fourth annual report represents the work of a large number of scientists from 8 Federal agencies, 2 State departments of agriculture, 22 universities, and several private research efforts.

In response to unexplained losses of U.S. honey bee colonies that began to be reported in 2006 as a condition named Colony Collapse Disorder (CCD), USDA's Agricultural Research Service (ARS) and National Institute of Food and Agriculture (NIFA) led a collaborative effort to define an approach for characterizing and resolving the problem; this effort resulted in the CCD Action Plan in July 2007. Many universities and Federal, State, and private organizations helped develop the plan and, through 2011, continued to carry out work to address CCD.

A survey of managed honey bee colonies found overall losses of 22 percent on average during winter 2012 (October 2011 to April 2012), down from losses of approximately 33 percent reported in 2010. In general, in the years since CCD began to be reported, winter losses have been averaging around 33 percent, of which approximately one-third was attributed to CCD. Many factors could be involved in the lower level of overall losses in 2011, including the mild winter weather in many parts of the United States. Nevertheless, the overall proportion of winter losses is still high, and the continued economic viability of pollination by honey bees remains threatened.

Although a number of factors continue to be associated with CCD, including parasites and pathogens, poor nutrition, pesticides, bee management practices, habitat fragmentation, and agricultural practices, no single factor or pattern of factors has been proven to be "the cause" of CCD.

Some recent scientific evidence appears to indicate that some pesticides may have sublethal effects on honey bees; however, the relevancy of some of these effects to estimating overall risks to honey bees is uncertain. In early 2010, ARS held a workshop with the U.S. Environmental Protection Agency to discuss how the potential effects of pesticides could be better documented in laboratory and field-based studies. ARS scientists and others have been examining whether pesticide-related effects may correlate with CCD incidents or other bee health problems.

The CCD Steering Committee is examining the current accumulation of knowledge concerning CCD and has begun the process of identifying new and relevant research needs to determine the cause or causes of CCD and how other factors or specific combinations of factors contribute to declining pollinator health. Scientists, beekeepers, growers, and other stakeholders have been invited to convene in October 2012 to update the CCD Action Plan for the next 5 years (2013–2017). Participants in the October meeting will consider the development of strategies to address the multiple factors (parasites and pathogens, poor nutrition, pesticides, and bee management practices) associated with diminished pollinator health in general, rather than focus exclusively on honey bees and CCD.

Research continued in 2011 in the four topic areas outlined in the 2007 CCD Action Plan. The studies encompassed honey bee and pollinator losses in general, not just CCD. Progress in the four topic areas is highlighted below.

- Topic I: Survey and Sample Data Collection. Surveys continued to provide evidence of high honey bee losses due to a combination of CCD and declines in pollinator health. Research has shown that weak colonies had overall higher pathogen levels and evidence of pesticide residues, although no pattern of specific pathogens or pesticides was indicated. Colonies in comparatively good health also contained a wide range of pesticide residues.
- Topic II: Analysis of Existing Samples. Previously, viruses and other pathogens and parasites were found to be present at greater levels in colonies affected by CCD than in those not affected by CCD. Studies in 2010 revealed several new viruses and other pathogens affecting honey bees. Additional studies to determine whether these new pathogens are involved in CCD-affected hives are continuing.
- Topic III: Research to Identify Factors Affecting Honey Bee Health, Including Attempts to Recreate CCD Symptomology. CCD was initially characterized by the rapid loss of adult worker bees from the colony, lack of dead worker bees in or immediately outside of the hive, and delayed invasion of hive pests. Researchers have observed that the Varroa mite and other pathogens such as Nosema may be contributing factors to CCD, although these occur at levels that are typically considered below economic damage thresholds. Researchers continued their efforts to document whether correlations exist between overall colony health and the presence of Varroa mites, diverse pathogens, and pesticides.
- Topic IV: Mitigation and Management Preventive Measures. Two national multi-year projects, the ARS Area-wide Project on Honey Bee Health and a NIFA-funded Coordinated Agricultural Project, progressed in developing management strategies to combat bee losses. The "Bee Informed Partnership" (http://beeinformed.org), sponsored by NIFA, continued to examine bee management practices and facilitate communication among beekeepers of practices that reduce colony losses while promoting bee health. The eXtension Community of Practice (www.extension.org/bee health) also continued to disseminate information on honey bee health and management practices. A National stakeholders' meeting scheduled for October 15–17, 2012, is intended to identify management practices for the different factors affecting honey bee health.

The Natural Resources Conservation Service has revised its appropriate conservation practices to encourage landowners to provide quality pollinator habitat and protect pollinators.

Chapter 6 - Mandated by section 7204(h)(4) of the Food Conservation and Energy Act, Pub. Law No. 110-246, the "2008 Farm Bill," this third annual report of research progress during 2010 on Honey Bee Colony Collapse Disorder (CCD) represents the work of a large number of scientists from 8 Federal agencies, 2 State departments of agriculture, 22 universities, and several private research efforts.

In response to the unexplained losses of U.S. honey bee colonies now known as Colony Collapse Disorder (CCD), USDA's Agricultural Research Service (ARS) and National Institute of Food and Agriculture (NIFA) led a collaborative effort to define an approach for responding to CCD, resulting in the CCD Action Plan, in July 2007. Other Federal agencies e.g., the Animal and Plant Health Inspection Service (APHIS), the Natural Resources Conservation Service, and the Environmental Protection Agency's (EPA's) Office of Pesticide Programs have been members of the CCD Steering Committee for many years; this year, USDA's Office of Pest Management Policy and the National Agricultural Statistics Service have joined the Steering Committee. Many universities and organizations (i.e., Federal, State, and private) were also involved in developing this plan and are carrying out work that addresses the CCD problem.

Annual surveys clearly show that overall colony losses continue to be as high as 30 percent or more since CCD began to be reported. Beekeepers cannot economically sustain such high levels of losses indefinitely. Although a number of factors have been associated with CCD and pollinator declines in general, no single factor or specific combination of factors has been identified as a "cause." Factors associated with declines include disease/parasites, nutrition, pesticides, bee management practices, habitat fragmentation, and agricultural practices. Reducing the incidence of CCD and pollinator declines will likely require managing multiple factors simultaneously. Overall losses to managed pollinators were about 33 percent. Relative to the overall losses, CCD contributed approximately 8 percent in recent national surveys. However, losses to individual beekeepers attributed to CCD may vary significantly by size of the beekeeping operation. Therefore, this statistic is just a rough estimate.

To help ensure more robust pollinator health, the CCD Steering Committee is facilitating the development of science-based best management strategies that integrate current research results associated with CCD incidence and pollinator decline in general. As part of these efforts, the CCD Steering Committee member agencies are working together to examine ways to mitigate impacts on pollinator health. This effort requires coordination across different Federal partners (USDA, Department of Interior, and EPA), States, and stakeholders to ensure a sustained effort that effectively integrates management options.

The CCD Steering Committee believes that it is necessary for scientists, beekeepers, and growers to reconvene in the next year to revisit approaches and research, which are now needed to counter honey bee decline, taking into account information developed during the past 3 years and to revise the 2007 Action Plan to reflect the current state of knowledge. A range of stakeholders will be engaged to develop strategies for managing factors associated with diminished pollinator health and not focus exclusively on CCD.

Research is ongoing in the four topic areas outlined in the CCD Action Plan, but the studies encompass factors involved in bee losses in general, not just due to CCD. The four areas are (1) survey and sample data collection, (2) analysis of existing samples, (3) research to identify factors affecting honey bee health, including attempts to recreate CCD symptomology, and (4) mitigation and management preventive measures. Progress in each of the four topic areas is highlighted below.

- Topic I: Survey and Sample Data Collection. Surveys continue to provide evidence of high honey bee losses due to CCD and declines in pollinator health. Correlated with these high levels of losses, research has shown that weak colonies had overall increased pathogen levels and showed evidence of pesticide residues, although no pattern of specific pathogens or pesticides was indicated. Colonies in comparatively good health also contained a wide range of pesticide residues. To understand the potential role of pesticides in pollinator declines, representatives of the CCD Steering Committee (APHIS, ARS, and EPA) participated in a Society of Environmental Toxicology and Chemistry global Pellston conference. The purpose was to develop a risk assessment process for honey bees and non-Apis bees and to identify the exposure and effects data needed to inform that process.
- Topic II: Analysis of Existing Samples. Previously, viruses and other pathogens and parasites were found to be present at greater levels in CCD colonies than in non-CCD colonies. The extent to which pesticides are associated with CCD remains uncertain, and additional research is necessary. Studies in 2010 revealed several new viruses and other pathogens affecting honey bees. Further studies are needed to determine if these new pathogens are involved in CCD-affected hives.
- Topic III: Research to Identify Factors Affecting Honey Bee Health, Including Attempts to Recreate CCD Symptomology. CCD was initially characterized by the rapid loss of adult worker bees from the colony, lack of dead worker bees, and delayed invasion of hive pests. Although additional studies are needed, researchers have recently observed that Varroa mite and other pathogens such as Nosema may be contributing factors to CCD. In addition, the Varroa mite and other pathogens occur at levels that are typically considered below economic thresholds. Researchers continue their efforts to document whether there are correlations between the presence of Varroa mites, diverse pathogens, and pesticides, which appear to impact overall colony health. In addition, apiaries surrounded by intensively farmed landscapes were found to be associated with higher colony losses in the spring.
- Topic IV: Mitigation and Management Preventive Measures. Two national multiyear projects, the ARS Area-wide Project on Honey Bee Health and a NIFA-funded Coordinated Agricultural Project (CAP), are continuing to make progress in developing managements strategies to combat bee losses. A new CAP project funded by NIFA, the "Bee Informed Partnership" (http://beeinformed.org), has begun to examine bee management practices and facilitate communication of successful practices between beekeepers. The eXtension Community of Practice (www.extension.org/bee_health) is also disseminating information on honey bee health and management practices.

Chapter 7 - Mandated by the 2008 Farm Bill [Section 7204 (h) (4)], this second annual report on Honey Bee Colony Collapse Disorder (CCD) research progress represents the work of a large number of scientists from 8 Federal agencies, 2 state departments of agriculture, 22 universities, and several private research efforts.

In response to the unexplained losses of U.S. honey bee colonies, now known as colony collapse disorder (CCD), the U.S. Department of Agriculture's (USDA) Agricultural Research Service (ARS) and National Institute of Food and Agriculture (NIFA) led a collaborative effort to define an approach to CCD, resulting in the CCD Action Plan in July 2007.

Many universities and organizations—Federal, State, and private—were involved in developing this plan and are carrying out the work that addresses the CCD problem.

In: Bee Health: Factors, Analyses, and Research Progress ISBN: 978-1-62618-044-4
Editor: Roger Dunham

Chapter 1

BEE HEALTH: THE ROLE OF PESTICIDES*

Linda-Jo Schierow, Renée Johnson and M. Lynne Corn

SUMMARY

Bees, both commercially managed honey bees and wild bees, play an important role in global food production. In the United States, the value of honey bees only as commercial pollinators in U.S. food production is estimated at about \$15 billion to \$20 billion annually. The estimated value of other types of insect pollinators, including wild bees, to U.S. food production is not available. Given their importance to food production, many have expressed concern about whether a "pollinator crisis" has been occurring in recent decades. In the United States, commercial migratory beekeepers along the East Coast of the United States began reporting sharp declines in 2006 in their honey bee colonies. The U.S. Department of Agriculture (USDA) reports that overwinter colony losses from 2006 to 2011 averaged more than 32% annually. This issue remained legislatively active in the 110th Congress and resulted in increased funding for pollinator research, among other types of farm program support, as part of the 2008 farm bill (P.L. 110-246). Congressional interest in the health of honey bees and other pollinators has continued in the 112th Congress (e.g., H.R. 2381, H.R. 6083, and S. 3240) and may extend into the 113th Congress.

This report:

- Describes changes in managed and wild bee populations, given readily available data and information. It focuses on managed and wild bees only, and excludes other types of pollinators, including other insects, birds, and bats. Data on managed honey bees are limited, and do not provide a comprehensive view of changes in bee populations. Data for wild bee populations are even more limited.
- Provides a listing of the range of possible factors thought to be negatively affecting managed and wild bee populations. In addition to pesticides, other identified factors include bee pests and diseases, diet and nutrition, genetics,

---

* This is an edited, reformatted and augmented version of Congressional Research Service, Publication No. R42855, dated December 11, 2012.

habitat loss and other environmental stressors, and beekeeping management issues, as well as the possibility that bees are being negatively affected by cumulative, multiple exposures and/or the interactive effects of each of these factors.
- Briefly summarizes readily available scientific research and analysis regarding the potential role of pesticides among the factors affecting the health and well-being of bees, as well as the statutory authority and related regulatory activities of the U.S. Environmental Protection Agency (EPA) related to pesticide use.

A 2007 report by the National Research Council of the National Academy of Sciences, *Status of Pollinators in North America,* provides a more detailed scientific context for this report and may be consulted for more in depth understanding about bee health. That study concluded that many factors contribute to pollinator declines in North America, and CRS accedes to that conclusion.

Accordingly, the focus of this report on bee exposure to pesticides is not intended to imply that pesticides are any more important in influencing the health and wellness of bees than any of the other identified factors influencing bee health. Pesticides are only one of the many influences on bee health.

Because neonicotinoid pesticides have been the focus of concerns in Europe and in the United States, this report briefly describes recent scientific research related to possible effects of exposure to these pesticides on bees.

The report concludes with a summary of recent regulatory activity regarding neonicotinoids at EPA, the federal agency charged with assessing risks and regulating U.S. sale and use of pesticides.

INTRODUCTION

Bees, both commercially managed honey bees and wild unmanaged bees, as well as many other types of insect pollinators, play an important role in global food production. In the United States, the value of honey bees only as commercial pollinators in U.S. food production is estimated at about $15 billion to $20 billion annually. The estimated value of other types of insect pollinators to U.S. food production is not available. Worldwide, the contribution of all insect pollinators—including commercial bees and wild bees—to the worldwide crop production used directly for human food is valued at €153 billion (or about $200 billion), according to the United Nations (UN).[1]

Given their importance to food production, many have expressed concern about whether a “pollinator crisis” has been occurring in recent decades. Worldwide reports indicate that populations of both managed honey bees and native bees have been declining, with colony losses in some cases described as severe or unusual. In Europe, managed honey bee colony numbers have been declining since the mid-1960s, and individual beekeepers have reported “unusual weakening and mortality in colonies,” particularly during the period spanning winter through spring.[2]

According to the United Nations, many insect pollinator species may be becoming rarer, causing some to question whether this is further sign of an overall global biodiversity decline.

In the United States, commercial migratory beekeepers along the East Coast of the United States began reporting sharp declines in their honey bee colonies starting in 2006. The U.S. Department of Agriculture (USDA) reports that overwinter colony losses from 2006 to 2011 averaged more than 32% annually.

This issue was legislatively active in the 110th Congress and resulted in increased funding for pollinator research, among other types of farm program support, as part of the 2008 farm bill (P.L. 110-246).[3] Congressional interest in the health of honey bees and other pollinators has continued in the 112th (e.g., see H.R. 2381, H.R. 6083, and S. 3240) and may extend into the 113th Congress.

Reasons cited for U.S. bee population decreases include a wide range of possible factors thought to be negatively affecting pollinator species. For managed and wild bee populations, potential identified causes for these declines include bee pests and diseases, diet and nutrition, genetics, habitat loss and other environmental stressors, agricultural pesticides, and beekeeping management issues, as well as the possibility that bees are being negatively affected by cumulative, multiple exposures and/or the interactive effects of each of these factors.

A 2007 report by the National Research Council of the National Academy of Sciences, *Status of Pollinators in North America*[4] (hereafter referred to as the "2007 NRC study") describes many of the factors affecting bee health and population effects.

Aside from the range of potential identified causes for bee colony declines, this report addresses what role, if any, pesticides play in influencing the health and wellness of bees. It begins by describing what is known about managed and wild bee populations, continues with brief descriptions of factors that might be affecting those populations, including a more detailed look at the potential effects of pesticide exposures, and concludes with a synopsis of pesticide risk management at the U.S. Environmental Protection Agency (EPA).

BACKGROUND

More than 200,000 species of insects (bees, butterflies, moths, flies, wasps, beetles, ants), bats, birds and hummingbirds, and small mammals are plant pollinators. This report focuses on two categories of bees only:

- ***Managed Honey Bees.*** Honey bees (*Apis mellifera*; Family: Apidae) are the most well-known bee species. Honey bees are considered to be "social" bees in that they tend to work together in a highly structured social order, consisting of cooperation and division of labor within a colony, as well as the presence of two generations in a single nest at the same time. This same social behavior also allows for bees to be domesticated and managed. However, honey bees are only one of the world's estimated 17,000 described bee species, and one of the estimated total of 20,000 to 30,000 bee species worldwide.[5] Honey bees are not native to North America, but were introduced by European settlers in the 1600s.
- ***Wild Bee Species.*** An estimated 4,000 species of bees are native to North America.[6] With few exceptions, most of these are wild and not managed bees. Most types of wild bees are "solitary" bees—estimated at about 3,500 species—and do not have long-lived colonies.[7] Data and information on native bees are more limited than information on managed bees.

  The five most common families of native bees in North America are Andrenidae, Apidae, Colletidae, Halictidae, and Megachilidae. Andrenid bees are all ground nesters, and mostly comprise a large family of dark, non-descript bees, although some are colorful. Bees in the large Apidae family include not only honey bees, but also bumble bees (such as *Bombus* spp.), carpenter bees, squash or gourd bees, and others. Bees in the smaller Colletidae family are broader and wasp-like, and include plasterer bees. Bees in the Halictidae family include many species of sweat bees, and most have shiny metallic colored or black bodies. Bees in the Megachilidae family include resin and mason bees, orchard bees, and leaf-cutter bees that mostly nest in holes. Most families of bees contain some types of parasitic and cuckoo[8] bees. For a summary see text box below.

Both managed and wild bee species are critical to plant pollination and are economically valuable to U.S. agricultural production. In addition to honey bees, other types of bees are also managed, such as bumble bees, orchard bees, and alfalfa leaf-cutting bees. Some of these bees exhibit some of the social behaviors commonly associated with honey bees. These include bumble bees and some types of stingless bees.

Only data on managed honey bees are readily available. Data and information on native bees are more limited. Where a few bee species are kept commercially and managed for their pollination services, wild bee species also play an important role in providing pollinator services to both commercial and small-scale home gardening systems.

Bee pollination of agricultural crops is said to contribute to the production of as many as 90 agricultural crops, including a wide range of high-value fruits, vegetables, tree nuts, forage crops, some field crops, and other specialty crops.[9]

A number of agricultural crops are almost totally (90%-100%) dependent on animal pollination, including apples, avocados, blueberries, cranberries, cherries, kiwi fruit, macadamia nuts, asparagus, broccoli, carrots, cauliflower, celery, cucumbers, onions, legume seeds, pumpkins, squash, and sunflowers.[10]

Other specialty crops also rely on pollination, but to a lesser degree. USDA reports that native bees also provide pollination services for a number of food crops. In addition to some of the aforementioned crops, these include alfalfa seeds, almonds, canola, chokecherries, grapefruit, pears, plums, prunes, soybeans (hybrid seed production), tomatoes, vegetable seeds, and watermelons.[11]

The economic value of pollination services provided by managed honey bees and wild bees is difficult to quantify.

One widely cited estimate of the monetary value of honey bees as commercial pollinators in the United States is $14.6 billion annually.[12] Most of the estimated value is attributable to alfalfa (mostly for alfalfa hay), apples, almonds, citrus, cotton, and soybeans. The 2007 NRC study reports other estimated values for honey bee pollination ranging from $5.7 billion—$19.0 billion annually.

For native bees, estimates of the economic value of pollination services are not readily available.

However, a 2011 study by researchers at the University of California estimated that wild bee species add $0.9 billion to $2.4 billion per year in value to California's agriculture through pollination services.[13]

Classification of Bees

Kingdom Animalia—Animal
Phylum Arthropoda—Arthropods
Class Insecta—insects, hexapoda
Order Hymenoptera—ants, bees, wasps
Superfamily Apoidea—bees, sphecoid wasps, apoid wasps
Family Andrenidae—andrenid bees, andrenids
Family Apidae—honey bees, bumble bees, carpenter bees, squash bees, stingless bees
Family Colletidae—colletid bees, plasterer bees, yellow-faced bees
Family Halictidae—halictid bees, sweat bees
Family Megachilidae—leafcutting bees, orchard bees, mason bees, resin bees
Family Melittidae—melittid bees, melittids

Source: Integrated Taxonomic Information System (ITIS), ITIS is supported by a consortium of federal and international agencies as well as scientific organizations to provide authoritative taxonomic information on known plant and animal species. Available at http://www.itis.gov/. Condensed and slightly modified by CRS to include more common names.

CHANGES IN BEE POPULATIONS

Managed Honey Bees

There are an estimated 115,000-125,000 beekeepers in the United States.[14] Most of these (roughly 90,000-100,000) are hobbyists with fewer than 25 hives.[15] Commercial beekeepers tend to have more than 300 hives, and migrate their colonies during the year to provide pollination services to farmers. In the United States, most pollination services are provided by commercial beekeepers.[16]

USDA does not compile comprehensive annual survey data on honey bee colonies. Available data are limited and not ideal for evaluating population changes among honey bees. Annual data on the number of honey bee colonies are from statistics tracking the number of operations that produce honey for commercial sale. Additional data on all "colonies of bees" are from USDA's five-year *Census of Agriculture* (the most recent available data are for 2007 and 2002). These data are not strictly comparable: Data are compiled for different purposes—the latter tracks all farms with bee colonies, the former tracks operations that produce honey for commercial sale; also data are compiled using different data collection techniques—one is a periodic census, the other a statistical estimate. Generally, USDA data are mostly intended to track the number of honey-producing colonies; however, it is the value of managed bees as crop pollinators that provides perhaps the greatest economic impact in the production of food and feed crops.

Data from the most recent USDA *Census of Agriculture* indicate that 2.9 million bee colonies were on U.S. farms in 2007.[17] This compares to data from the previous *Census* in 2002 when there were about 2.4 million bee colonies, reflecting an increase in the total number of bee colonies. However, using similar data for the 1950s-1960s, there were an estimated 6 million managed honey bee colonies.[18] This may reflect ongoing consolidation and structural shifts in the U.S. agricultural sectors, rather than conclusive trends in species populations, abundance, and distribution.

USDA annual data for honey-producing colonies are similarly inconclusive. These data show year-to-year fluctuations, but not a clear downward trend over time (Table 1). Over the 10-year period shown, the number of bee colonies dropped from 2.6 million bee colonies in 2002 to 2.5 million bee colonies in 2011.[19] However, the endpoints mask a 10-year high in 2010 when honey production increased and the number of honey bee colonies reached nearly 2.7 million (Table 1). This increase in the number of honey-producing hives may reflect efforts among beekeepers to continually replenish their hives—either in response to a colony die-off or increase in the number of hives to raise overall honey production or to provide additional pollinator services. This increase might also reflect the rising popularity of beekeeping as a popular part-time hobby.

Table 1. Honey-Producing Bee Colonies, 2002-2011

Year	Colonies (1,000 colonies)	Honey Production (1,000 pounds)
2002	2,574	171,718
2003	2,599	181,727
2004	2,556	183,582
2005	2,413	174,818
2006	2,393	154,907
2007	2,443	148,341
2008	2,342	163,789
2009	2,498	146,416
2010	2,692	176,462
2011	2,491	148,357

Source: USDA, *Honey,* http://usda.mannlib.cornell.edu/MannUsda/view Document Info.do? documentID=1191. Honey producing colonies for producers with 5 or more colonies. Areas with the most honey bee colonies include North and South Dakota; California; Florida; Montana, Minnesota, Idaho, and Texas. Other major states are Michigan, Oregon, Georgia, Nebraska, New York, Washington, Wisconsin, and Wyoming.

Among all honey bee colonies, more than 2 million are reported to belong to commercial migratory beekeepers and are rented each year to pollinate U.S. agricultural crops. An estimated 1.5 million colonies are needed each year to pollinate California's 750,000 acres of almond trees alone.[20] Increasingly bee colony rentals are being used to pollinate cultivated blueberries in the East Coast states and in the Midwest.[21] Other information indicates that bee colonies are also rented for apple, pear, plum, cherry, cranberry, avocado, cucumber, kiwi fruit, melon, pepper, and citrus fruit production, as well as for alfalfa, clover seed, and sunflower production.[22] Both locally and globally, some are concerned that the availability of honey bee stocks is not keeping pace with growing agricultural demands for pollination services.[23]

The 2007 NRC study voices concerns about the available USDA data on honey bees, cautioning against use of these data to determine changes in honey bee populations. Concerns about these data, including: the purpose of these data is to track U.S. honey production not bee populations; lack of comparability among the available data compilations; the potential for misidentification of species and miscounting because of data collection procedures; and other issues.[24]

Notwithstanding these concerns about data availability and quality, the 2007 NRC study concluded: "Long-term population trends for the honey bee, the most important managed pollinator, are demonstrably downward."[25] Honey bee colony losses are not uncommon. The 2007 NRC study cites USDA data showing honey bee declines in 1947–1972, and 1989–1996, as well as declines starting in 2005 (despite reports of a sharp rise in 2010). These downward trends have continued since that study was published, attributable to what is now known as Colony Collapse Disorder (CCD), which is thought to have affected mostly managed bee colonies starting in late 2006.

As reported by USDA: "The defining characteristic of CCD is the disappearance of most, if not all, of the adult honey bees in a colony, leaving behind honey and brood (immature bees confined to cells in the hive, including larvae and pupae) but no dead bee bodies."[26] Scientists at USDA and bee labs across the country have been looking for the cause or causes of CCD within four broad categories:[27]

- Pathogens (such as *Nosema ceranae*);
- Parasites (such as *Varroa* mites);
- Environmental stressors (such as pesticides or lack of nectar diversity); and also
- Management stressors (such as transportation stress by migratory beekeepers).

As reported by USDA in July 2012: "While many possible causes for CCD have been proposed, reported, and discussed—both in the scientific literature and popular media—no cause has been proven."[28] These and other factors are discussed in "Identified Factors Affecting Bee Health".

Heightened attention following concerns about CCD has resulted in better tracking by USDA of annual honey bee colony losses. USDA reports that before 2006 bee colony losses averaged 15%- 20% per year since the 1990s, attributable to a variety of factors, such as mites, diseases, and management stress. By comparison, bee colony losses between the winters of 2006/2007 and 2010/2011 averaged more than 32% during the year: Surveys found total colony losses of:[29]

- 30% in the winter of 2010/2011;
- 34% in the winter of 2009/2010;
- 29% in the winter of 2008/2009;
- 36% in the winter of 2007/2008;
- 32% in the winter of 2006/2007.

Other information from USDA indicates that colony losses range from 7% to 80% loss, depending on area.[30] Preliminary survey results indicate that total losses of managed honey bee colonies from all causes were 22% nationwide for the 2011/2012 winter, representing a substantial improvement in mortality compared to the previous five years.[31]

The leading causes of declining bee populations cited in the 2007 NRC study—which was published before CCD became a well-publicized phenomenon—included pathogens and introduced parasites, particularly *Varroa destructor*, the varroa mite. That study, among others, documents the extensive, but still inconclusive literature on honey bee population losses due to bee pests, parasites, pathogens, and disease, as well as other causes.

Native Bees

While managed bees are commercially important, wild bee species are important ecologically for sustainable forests and fields. No comprehensive formal statistics are available on populations of wild bees in the United States, despite numerous attempts to promote the use of various informal "citizen science" initiatives in California, Florida, other East Coast states, and elsewhere.[32] As already noted, an estimated 3,500 species are solitary bees, and these species are not readily surveyed. Roughly an additional 500 species of bees native to North America show some degree of social behavior. Scientific literature is replete with assertions about the paucity of data on one group or another of native bees. Long term data are particularly difficult to obtain for wild bees. As concluded by the researchers in the 2007 NRC study, among all wild pollinators:

> There is evidence of decline in the abundance of some pollinators, but the strength of this evidence varies among taxa. Long-term population trends for several wild bee species (notably bumble bees) ... are notably downward. For most pollinator species, however, the paucity of longterm population data and the incomplete knowledge of even basic taxonomy and ecology make definitive assessment of status exceedingly difficult.[33]

In the absence of comprehensive data, the following section describes selected information gained about particular taxa or sites since the publication of the 2007 NRC study. As some of the studies cited below demonstrate, scientists seeking to find any data to analyze broad pollinator trends might be forced to rely on amalgamations of disparate studies, collections by citizen scientists,[34] and other unconventional approaches.

Timing of Bees and Associated Flowering Plants

A 2011 paper showed that the phenological changes in flowering times were accompanied by changes in the spring emergence of bee populations.[35] Reviewing a number of previous studies on flowering times and pollination, the investigators found that ten species of bees (including both solitary and social species) had advanced to earlier dates in the spring for their first emergence, and the flowering times of the plants they visited also advanced. Advances in emergence time over the period from 1970-2010 were highly significant. Various hypotheses to explain the difference were examined; the data "[support] the idea that climate change is the main factor explaining the observed phenological advances." However, the authors "conclude that phenological mismatch probably has not occurred already, but that it could occur in the future, as rates of temperature warming increase." At the same time, the study noted that it analyzed generalist bees and generalist flowers—bees that visit many plant species, and plants visited by many bee species.[36] Where the life cycles of particular bee species are tied to particular plant species, different trends may occur. While no data were reported on the population sizes of these bees, in this study there was no clear connection between climate change (as measured by flowering times) and any harm to the species studied. The bees seemed to be keeping up with the changes in dates of flower availability.

In 2010, an international group of scientists reviewed global trends in pollinator species, including bees.[37] They cited a study based on data gathered by citizen-scientists in the United Kingdom and in the Netherlands (number of years of observation varying by location) showing that "in both countries, bee diversity has fallen significantly in most landscapes.... Analysis of pollinator traits demonstrated that in bees ..., specialist (diet and/or habitat) and sedentary species tended to decline, whereas mobile generalists tended to thrive." As the authors also noted:[38]

> The parallel dynamics between plants and their pollinators suggest a link between the two, although this is correlative and the mechanism is as yet unknown. It could be that plant declines are caused, in part, by a lack of pollination services, or bees could be declining owing to a lack of floral resources, or indeed both could be declining owing to shared sensitivity to environmental changes.

Bumble Bees

Compared to most native species, bumble bees (*Bombus)* are better studied. According to the 2007 NRC report, there are 239 species of bumble bees worldwide; of these, 60 species are found in the U.S., Mexico, and Canada.[39] Some species are managed in controlled environments to pollinate greenhouse tomatoes. Others are valued as primary or supplementary pollinators for members of the squash family such as cucumbers, watermelons, and cantaloupes.

In the United States, citizens are recruited to gather data on these large insects—for example, forming Bumble Bee Brigades in one campaign.[40] Even so, information about the wild bumble bees is not comprehensive. One species (*B. franklini*) is thought to be extinct in its very limited (former) range in northeastern California and southeastern Oregon. Hoping to establish trends for some species, some scientists examined both museum collection records and intensive nationwide surveys for records of abundance and distribution of eight species of this genus.[41] Four species have declined in relative abundance by up to 96%. These four also contracted their ranges by 23% to 87%. These changes were statistically highly significant; the other four species showed no clear pattern in terms of range reduction. The authors hypothesized that the four declining species may be affected by a pathogen (*Nosemi bombi*) as well as reduced genetic diversity.[42] The data were adequate only to show an association, rather than a cause—were the four species declining because of the disease and low genetic diversity? Or was some other factor causing these species to decline and making them more susceptible to the disease and also reducing their genetic diversity?

Stingless Bees

Meliponine bees, with about 400 species, belong to the same family (Apidae) as bumble bees and honey bees.[43] Commonly called *stingless bees*, they are found in the tropics, and were widely cultivated by Native Americans in Central and South America before the introduction of honey bees from Europe. While they do not sting, most species can inflict a serious bite. A few species can produce a secretion from their mouth parts; added to a bite, the secretion may produce blisters. Loss of forest cover eliminates nesting sites. The shift by beekeepers to honey bees for more intensive production has also reduced the number of stingless bees. An analysis of the effects of forestry on stingless bees in Brazil observed that most bee nests were found in hollow trees over 50 cm (~20 inches) in diameter.[44] Given the high value of the bees as pollinators for many species of plants, and the low commercial value of hollow trees, the authors recommended that managed forests maintain hollow trees as both seed sources and as homes for these bees.[45] In addition, some studies suggest that stingless bees are especially susceptible to some pesticides.[46]

Interactions of Bees and Associated Flowers: Trophic Cascade

The phenomenon known as *trophic cascade* may also play a role in the interaction of pollinator species. Many plants can be pollinated by more than one species, and many bee species use more than one pollen or nectar source. As individual pollinator species (bees or other species) decline, other species may take over the declining species' role. Where the data are available (as with the *Bombus* study cited above), evidence suggests that, where a generalist bee species is able to step in to provide pollination services (even if less able than a specialist bee), the effects of a reduced bee population may be masked by the generalist. In more concrete terms, the presence of a highly efficient generalist—like the honey bee—may mask the loss of native species (regardless of the reason for the decline of the native species). If honey bees themselves then decline, the effects of the lost population may be more severe because populations of back-up pollinators have already been reduced or eliminated. And conversely, reductions in honey bee populations may reduce competition sufficiently to allow other (reduced) bee populations to rebound to greater numbers.

IDENTIFIED FACTORS AFFECTING BEE HEALTH

Factors that have been identified to be negatively affecting managed and native bee populations are (listed in no particular order):[47]

- **parasites, pathogens, and diseases** (for a detailed listing, see text box);
- **bee genetics** including lack of genetic diversity and lineage of bees, and increased susceptibility and lowered disease resistance, and also miticide resistance by the mites;[48]
- **diet and nutrition** including poor nutrition due to apiary overcrowding, pollination of crops with low nutritional value, and pollen or nectar scarcity associated with invasive plants;
- **bee management issues** including transportation stress from migratory beekeeping, overcrowding, feeding practices, chemicals used by beekeepers to control mites (antibiotics and miticides), confinement and temperature fluctuations, susceptibility to disease, potential for cumulative exposure to diseases and parasites, use of bees for honey production versus pollination, chemical residue or contamination in the wax, and reliability of the queen source;

- **habitat loss, and other environmental or biological stressors** including loss of foraging area, interspecific competition between native and non-native bees, pathogen spillover effects, and climate change;
- **pesticides** including acute or cumulative exposure to new types and combinations of agricultural pesticides through a variety of media including dust, water droplets, pollen, and nectar;
- **other agricultural practices** including the use of genetically incorporated pesticides in seeds or treated seeds, such as with bioengineered crops; and
- **potential cumulative and interactive effects** of each of these factors.

These same factors were among those initially thought to be contributing to CCD. As outlined in USDA's 2011 progress report, the available research over the past few years on the numerous possible causes for CCD has led USDA and university researchers to conclude that "no single factor or specific combination of factors has been identified as a 'cause' [for CCD]."[49] This has led researchers to further examine the hypothesis that CCD may be "a syndrome caused by many different factors, working in combination or synergistically;"[50] it may also involve "an interaction between pathogens and other stress factors."[51] USDA reports that researchers are continuing to focus in three major areas: (1) pesticides that may be having unexpected negative effects on honey bees; (2) a parasite or pathogen that may be attacking honey bees, including the pathogenic gut fungus, *Nosema*, as well as viral infections; and (3) a combination of existing stresses that may compromise the immune system of bees and disrupt their social system, making colonies more susceptible to disease and collapse.[52]

Pests and Diseases Affecting Honey Bees

As noted in the 2007 NRC study, among the leading causes of managed honey bee losses are diseases, parasites, and recently introduced competitors. Most notable are declines due to two parasitic mites, the so-called vampire mite (*Varroa destructor*) and the tracheal mite (*Acarapis woodi*). Also of concern is the emergence of new or newly virulent fungal and viral diseases. New invasive pests are also harming bees. Below is a brief listing of some identified pests and diseases. For more information, see the 2007 NRC report and other beekeeper guidance documents. The following is mainly excerpted from *The Beekeepers Handbook*. Honey bee diseases may be caused by a protozoan (e.g., bacteria or amoeba), fungus, or virus.

- *Nosema* disease is the most common adult bee disease, and is caused by a microscopic fungus (formerly considered to be a protozoan). Two *Nosema* species are found in honey bees: (1) *Nosema apis*; and (2) *Nosema ceranae* (the more virulent of the two).
- American Foulbrood Disease (AFD) is the most destructive of brood diseases. (Brood refers to the egg, larval, and pupal stages in bee development.) AFD is caused by a bacterium (*Paenibacillus larvae*) that exists in a spore or a vegetative stage; the disease is transmitted by the spore and the infected brood is killed by the vegetative stage.
- European Foulbrood Disease (EFD) is also caused by a bacterium

(*Melissococcus plutonius*) and is commonly found in colonies already weakened by lack of food or by other stressors.

- Chalkbrood disease is caused by a fungus (*Ascophaera apis*); adult bees can detect and remove diseased larvae; honey production may be reduced but the disease usually will not destroy a colony.
- Amoeba disease is caused by the amoeba *Malpighamoeba mellificae* Prell; it infests the gut of honey bees. Resistant spores (cysts) form in honey bees and can transmit the disease to other bees.
- Other less common adult diseases include septicemia and spiroplasma. Both are caused by bacteria. The former cause destruction of connective tissues; the latter infects the bees' blood. They tend to cause dysentery, and arise primarily from poor food and long periods of confinement.
- Some common viral diseases affecting honey bees include: Deformed Wing Virus (DWV); Black Queen Cell Virus (BQCV); Israeli Acute Paralysis Virus (IAPV); Acute Bee Paralysis Virus (ABPV); Sacbrood Virus (SBV); Kashmir Bee Virus; and Chronic Bee Paralysis Virus (CBPV).

Other pests are mostly invertebrates, but some vertebrates are problematic locally. Among these pests are parasitic mites, insects, and some larger animals.

- Tracheal mites (*Acarapis woodi*) are parasitic mites that live inside the breathing organs of adult bees, and eventually a newly mated female mite emerges from the old host bee.
- Varroa mites, now known mostly for the vampire mite (*Varroa destructor*), are large mites that feed on infested bees, resulting in disfigured, stunted adult bees and deformed larvae and pupae (varroosis).
- Emerging threats such as the parasitic phorid fly (*Apocephalus borealis*), known to parasitize bumble bees, has been found to also parasitize honey bees and can eventually cause bees to abandon their hives.
- Major insect enemies include the wax moth, and the small hive beetle.
- Minor insect enemies include assassin and ambush bugs, robber flies, mantids, wasps, and dragonflies.
- Various vertebrate pests include skunks, raccoons, bears, and mice.

One competitor deserves mention—the introduced Africanized honey bee. This bee is also a honey bee (*Apis mellifera*) but of a different strain from those imported from Europe. They are good foragers, but the colonies are more difficult to manage than the domestic European bee; they are sometimes called "killer bees" although an individual bee's sting is no more severe than that of a European bee. They were accidentally released in Brazil decades ago and have spread into the southern United States.

Sources: D. Sammataro and A. Avitabile, The Beekeepers Handbook, 4th Edition, pp. 189-232; Beekeepers Association of Northern Virginia, "Diseases, Parasites, Pests, and Predators," presentation materials; and for phorid flies, see A. Core, et al., "A New Threat to Honey Bees, the Parasitic Phorid Fly *Apocephalus borealis*," PLoS ONE 7(1), January 3, 2012, http://www.plosone.org/article/info:doi%2F10.1371%2Fjournal.pone.0029639.

POSSIBLE ROLE OF PESTICIDES

The previous section lists a range of possible factors thought to be negatively affecting managed and wild bee populations. Among the identified factors were bee pests and diseases, diet and nutrition, genetics, habitat loss and other environmental stressors, beekeeper issues, and pesticides, as well as possible negative effects of cumulative, multiple exposures and/or the interactive effects of each of these factors.

In response to congressional requests, this section examines in greater detail the potential role of pesticides, providing a summary of selected, readily available data and reviews of the scientific literature. This focus on the potential role of pesticides is not intended to imply that pesticides are as important or more important in influencing the health and wellness of bee colonies as compared with other identified factors influencing bee health; the relative importance of pesticides in U.S. or global bee health is unknown and a subject of numerous research projects, some of which are discussed below. In addition, this section summarizes general information about the nature of pesticides, pesticide uses, and pesticide regulation in the United States, as well as more specific information about the registration status of a class of pesticides known as neonicotinoids.

Pesticides and Pesticide Law

Pesticides are broadly defined in U.S. law as chemicals and other products used to kill, repel, or control pests.[53] Familiar examples include pesticides used to kill insects (insecticides) and weeds (herbicides) that can reduce the yield, and sometimes harm the quality, of agricultural crops, ornamental plants, forests, wooden structures (e.g., through termite damage), and pastures. But the broad legal definition of "pesticide" also applies to products with less familiar "pesticidal uses." For example, substances are pesticides when used to control mites, mold, mildew, and other nuisance growths in hives or on equipment. The term also applies to disinfectants and sterilizing agents, animal repellents, rat poison, and many other substances. An estimated 18,000 pesticide products are currently in use in the United States.[54] Pesticides vary greatly in toxicity, persistence in the environment, and ability to bioaccumulate up the food chain, as well as in the range of plants and animals that are likely to be affected in the event of exposure. Some are practically nontoxic to some species but exquisitely toxic to other species.

Statutory Framework

All pesticides are regulated by EPA under the authority of the Federal Insecticide, Fungicide, and Rodenticide Act (FIFRA), but approximately 5,800 pesticide products used in food production also are regulated under the Federal Food, Drug, and Cosmetic Act (FFDCA), Section 408. FIFRA requires EPA to regulate the sale and use of pesticides in the United States through product registration and labeling. EPA is required to regulate so as to prevent unreasonable adverse effects on people and the environment, taking into account the costs and benefits of various pesticide uses. FIFRA prohibits sale of any pesticide in the United States unless it is registered (licensed) and labeled to indicate approved uses and restrictions. It is a violation of the law to use a pesticide in a manner that is inconsistent with the label instructions. EPA registers each pesticide product for each approved use. For example, a product may be registered for use on bee hives to control mites or as a seed treatment for corn. In addition, FIFRA requires EPA to reregister pesticides first registered prior to 1984 and to review all registered pesticides periodically on a 15-year cycle based on new data that meet current regulatory and scientific standards.

For the 600 or more pesticides (i.e., active ingredients) registered for use in food production, section 408 of the FFDCA authorizes EPA to establish maximum allowable residue levels (also known as "tolerances") to ensure that human exposure to the pesticide ingredients in food and animal feed will be "safe."[55] A "safe" tolerance is defined in the law as a level at which there is "a reasonable certainty of no harm" from the exposure, even when considering total cumulative and aggregate pesticide exposure of children. (This standard also is used in registering pesticides for residential uses.) Under the FFDCA, foods (or animal feeds) with a residue of a pesticide ingredient for which there is no tolerance established, or with a residue level exceeding an established tolerance limit, are declared "unsafe" and "adulterated"; such foods cannot be sold in interstate commerce or imported to the United States. Pesticides may not be registered under FIFRA for use on food crops unless tolerances (or exemptions) have been established under the FFDCA.

Registration Process

When pesticide manufacturers apply to register an "active" ingredient for a pesticide, a commercial pesticide product, or a new use of a registered pesticide under FIFRA Section 3, EPA requires them to submit scientific data on toxicity and behavior in the environment. EPA may require data from any combination of more than 100 different tests, depending on the potential toxicity of active and inert ingredients and degree of exposure. To register a pesticide for use on food, EPA also requires applicants to determine the amount of residue that could remain on crops, as well as on (or in) food products (such as corn syrup), assuming that the pesticide product is applied according to the manufacturers' recommended rates and methods. Based on the data submitted, EPA determines whether and under what conditions a proposed pesticide use would present an unreasonable risk to human health or the environment, and, for a food or residential use, whether its use would be safe. Some features of pesticides that might affect registration decisions include the specificity of the pesticide for the targeted pest, its toxicity to people who apply it, its tendency to persist in the environment over time, and its ability to bioaccumulate in animals higher in the food chain. EPA specifically takes into account unintended harm to bees and available information for other nontargeted insects in its registration decisions. EPA requires studies to determine acute (short-term) toxicity of a pesticide on individual bees when they come into body contact with pesticide residue. EPA also collects reports on bee-kill incidents. If a pesticide appears to be very toxic to bees, EPA may require long-term studies of its effects.

If the risk is determined to be unreasonable or unsafe, EPA attempts to mitigate the risk by adjusting requirements on the label (for example, requiring a buffer zone around lakes and streams or requiring personal protective equipment for pesticide handlers). If the risk remains unreasonable or unsafe, EPA will refuse to register the pesticide. If the risk is determined to be reasonable and safe, registration is granted and the agency specifies the approved uses and conditions of use, including safe methods of pesticide storage and disposal, which the registrant must explain on the product label. EPA can and often does require specific application methods to be printed on the product label to minimize environmental damage.

For example, the label sometimes requires that application of certain pesticides occur when bees are not foraging, when there is little wind, or in a granular form or as a seed coating rather than aerially, in order to minimize spray drift off property. Pesticide registrations are reviewed at least once every 15 years to consider new scientific information and may be reviewed at any time in response to reports of adverse effects and possible unreasonable risks from use of particular pesticides.

Exposure to Pesticides

Beekeepers use pesticides registered for the control of bacteria, fungi, mites, and other bee pests. These are meant to protect bees from diseases. These pesticides are applied within and in the vicinity of hives. Some common pesticides applied deliberately to hives include formic acid, thymol, coumaphos, and fluvalinate.

Bees also sometimes are exposed to pesticides accidentally, either when pesticides are misused or misapplied, or when used according to label directions to control pests in areas frequented by bees, for example, alongside roads or rights of way for the control of weeds, trees, or other pests; on or near commercial farm crops; or on or near fields, lawns, and gardens to control fleas, ticks, weeds, grubs, mosquitos, or other adult insects. Figure 1 illustrates some significant paths of exposure to pesticides applied as a spray or as a soil or seed treatment (systemic) for bees. If bees happen to fly through a newly treated field or dust clouds from planting of seeds coated with pesticide, or if they are orally exposed to pesticide in food or water, and if exposure is high enough, bees may be sickened or die from pesticide exposure. (See summaries of specific studies below.)

Impacts of Pesticides on Bee Health

As explained above, little is known about the health over time of native bees in the United States. The health of managed honey bees is better characterized, but still somewhat uncertain. Therefore, it is not appropriate at this time to conclude there has been a general and significant decline in native bees, or to attribute global or even North American declines in the health of any particular species of bee either native or managed bee species to pesticides. On the other hand, pesticides are known to have some adverse local impacts on honey bees and some native bees. Widespread use of herbicides reduces habitat available to bees;[56] many pesticides are known to be acutely toxic to bees, given sufficient levels of exposure; and some reports of local bee kill incidents have been well documented. Laboratory research has shown lethal and sub-lethal effects on individual bees associated with exposure to numerous pesticides. According to research cited by the NRC, "The application of pesticides, especially insecticides used to control crop pests, kills or weakens thousands of honey bee colonies in the United States each year."[57] Nevertheless, the NRC concluded in its 2007 report that *local* bee kills "likely have not contributed significantly to the recent *national* decline in colony populations" [emphasis added].[58]

Five years after the NRC issued its report, generalizations about the relative importance of pesticides for global bee health still cannot be drawn from available data, given the disparate study designs and results. Moreover, although research has been and continues to be conducted, most scientists have focused on a single pesticide or pathogen at a time and consideration of interactions has been minimal.

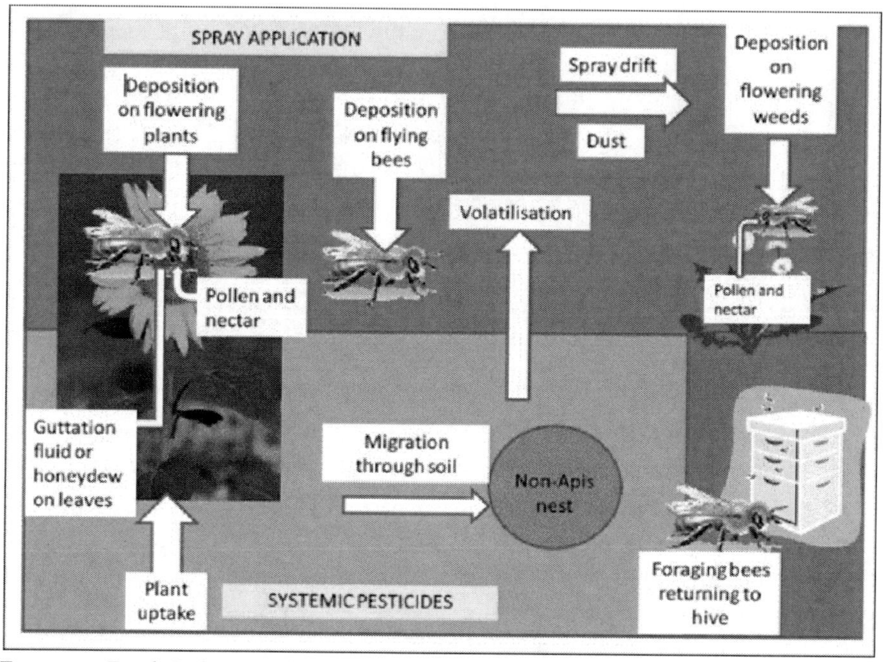

Source: European Food Safety Authority (EFSA), "Scientific Opinion on the science behind the development of a risk assessment of Plant Protection Products on bees (*Apis mellifera, Bombus* spp. and solitary bees)," Figure 3.1, EFSA-Q-2011-00417, May 2012, http://www.efsa.europa.eu/en /efsajournal/pub/2668.htm.

Figure 1. Major Routes of Exposure of Foraging Bees to Pesticides.

A 2010 article by a group of well-known researchers explains:

> Attempts to correlate global bee declines or CCD with increased pesticide exposures alone, have not been successful to date ... Pesticide interactions among various mixtures as well as with other stressors including *Varroa* and *Nosema*, IAPV, beneficial hive microbes, and impacts on bee immune systems all require further study. It seems to us that it is far too early to attempt to link or to dismiss pesticide impacts with CCD.[59]

Even with respect to pesticides known as neonicotinoids (see below), experts disagree about their potential effect on bee health; while some find that dietary neonicotinoids do not appear to be implicated in overall decline of honey bee populations,[60] others attribute unusual bee disappearances and deaths mostly to use of clothianidin (a neonicotinoid) on sunflowers and corn.[61] A 2010 research paper argues that experts and beekeepers may be arguing past each other, changing the definition of the problem depending on the audience. Some French researchers have concluded that at least some French stakeholders have intentionally employed changing definitions deliberately to confuse public understanding.[62]

This section summarizes selected recent studies that have raised questions about the risk to bees posed by chronic exposure to neonicotinoids and certain other pesticides, particularly during the spring planting season.

Generic Pesticide Research

Scientists have tested samples of beebread,[63] trapped pollen,[64] brood nest wax,[65] beeswax foundation,[66] and adult bees and brood and found a broad range of pesticides, including miticides, fungicides, insecticides, and herbicides.[67] Both healthy and unhealthy hives were examined, but no patterns were identified that might point to a particular pesticide or class of pesticides directly affecting bee health.

These same scientists noted that the doses of individual pesticides found in bees were not lethal, but researchers remained concerned about possible chronic problems caused by long-term exposure and possible additive or synergistic effects of exposure to the combinations of pesticides found.[68] Studies have found synergistic adverse effects on honeybees exposed chronically to sublethal concentrations of a pesticide (fipronil or imidacloprid) and *Nosema*.[69]

Pesticide products include ingredients other than the "active" ingredients which generally receive less scrutiny during registration. Some of these "inerts" are surfactants which are intended to improve delivery of the active ingredient to the target pest. A 2012 study looked at the impact on bee health of three categories of inerts, also known as adjuvants: nonionic surfactants, crop oil concentrates, and organosilicone surfactants.[70] Bee behavior was adversely affected after ingestion of 20 micrograms (μg) of organosilicone surfactant. Nonionic adjuvants also had a small effect, while the crop oil concentrates were inactive, according to the scientists.[71]

Research on Neonicotinoid Pesticides

Neonicotinoids comprise a class of pesticides (that is, active ingredients) that have come under considerable scrutiny with respect to their potential effects on bee health due to an incident of misuse (that is, use not in accord with the pesticide label) of one neonicotinoid, imidacloprid, in Germany[72] that resulted in a large bee kill and widespread beekeeper concerns about use of another neonicotinoid, clothianidin, and its impact on bees in France.[73] Neonicotinoids are systemic pesticides that, regardless of application method (spray, drip irrigation, granular spreading, or seed coating), once taken into the plant migrate into all parts, including flowers, pollen, and nectar.[74] In the United States, some of the most commonly applied neonicotinoids include imidacloprid, clothianidin, and thiamethoxam, which are sold by such tradenames as Admire®, Acceleron®, Axcess®, Attendant®, Belay®, Cruiser®, Gaucho®, Nitro Shield®, Poncho®, and Trimax Pro®. According to scientists at Bayer Crop Science, neonicotinoids accounted for almost 25% of the global pesticide market, and imidacloprid was the largest selling insecticide in the world in 2009.[75]

Exposure to Neonicotinoids

Bees can be exposed to neonicotinoids in many ways. A team of researchers recently reported that they had identified multiple routes of exposure to low levels of neonicotinoids for honey bees living and foraging near agricultural fields planted with corn or soybeans.[76] The highest potential exposure to the pesticides appeared to occur during planting season, when bee mortality was also high, according to the researchers. Clothianidin was found in about half the corn pollen they sampled, thiamethoxam in three of twenty samples, and fungicides in all pollen samples. Clothianidin was detected in all dead and dying bees but in no healthy bees. Although corn is not an insect-pollinated crop, this research demonstrated that bees forage for corn pollen and take it back to the hive.[77] About half the hive pollen sampled came from corn in this study. In addition, these scientists documented high levels of pesticide in exhaust material from mechanical planters when pesticides and talc were used to coat seeds, and found clothianidin in soil samples from planted and unplanted fields.[78]

Other studies have measured neonicotinoids in pollen and nectar of canola (rape), [79] corn,[80,81] and sunflowers[82,83] grown from seed coated with pesticides. Levels found were below those known to be acutely toxic. However, very recent studies have found that imidacloprid and thiamethoxam concentrations in nectar were greater in squash[84] and pumpkin flowers[85] when insecticide was applied to the soil than they were in canola and sunflowers grown from seed treated with neonicotinoids. Researchers also found metabolites of imidacloprid and thiamethoxam (clothianidin) in all parts of squash plants, along with the parent compound.[86] Krupke et al. found levels of neonicotinoids in bee-collected corn pollen that were similar to levels of imidacloprid determined by other scientists to have sublethal effects potentially affecting colony health.[87] (Some of these studies on the effects of neonicotinoid exposure are described below.)

Another potential source of exposure was suggested by Chengsheng Lu and colleagues, who showed that bees might be exposed to neonicotinoids in the corn syrup they are sometimes fed during the winter by beekeepers.[88] However, they did not sample corn syrup actually fed to bees, but rather showed that bees would consume sufficient imidacloprid to produce toxic sublethal effects if they were provided contaminated corn syrup in the hive. Research by DeGrandiHoffman, Sammataro, and Simonds found no pesticides in samples of high-fructose corn syrup obtained from three major suppliers.[89]

Effects on Bee Health

Neonicotinoids are insect neurotoxins that vary in strength of effect exerted on honey bees.[90] They are related to nicotine and were developed as an alternative to highly toxic (to humans) organophosphate insecticides such as methyl parathion.[91] Effects on individual bees may be lethal or sublethal depending on dose and other conditions of exposure. For example, EPA has determined that clothianidin "has the potential to be highly toxic on both a contact and an oral basis" to honey bees.[92] In addition, EPA has reported that one honey bee field study submitted to the agency indicates that "mortality, pollen foraging activity, and honey yield were negatively affected by residues of clothianidin," but the residue levels were not reported.[93]

Acute effects also have been demonstrated in another recent field study. It showed that honey bees can be killed by exposure to pesticide-contaminated talc if they fly through dust clouds associated with planting,[94] but mortality appears to depend on high levels of humidity.[95]

In response to reports that honey bees are disappearing and causing hives to collapse, recent studies of the impacts of exposure to imidacloprid and other neonicotinoids[96,97] have focused more on their potential to affect complex behaviors in insects, including flight, navigation, olfactory memory, recruitment, foraging, and coordination.[98] One study has reported sublethal effects of neonicotinoid pesticides on honey bee foraging behavior that may impair the navigational and foraging abilities of honey bees.[99] Scientists at Bayer Crop Science argue that the dose of thiamethoxam delivered to bees in this case was not "field-relevant."[100] At "field-relevant" doses for sunflowers and canola (defined as 0.15 nanogram (ng) per bee of imidacloprid or 0.05 ng/bee of clothianidin), Schneider et al. found no effect on foraging and homing behavior of honey bees exposed to treated crops.[101] The same study found, that "both substances led to a significant reduction of foraging activity and to longer foraging flights at doses of 1.5 ng/bee of imidacloprid or 0.5 ng/bee of clothianidin during the first three hours after treatment."[102] However, these latter numbers might be approaching "field relevant" numbers for neonicotinoids in squash nectar and corn pollen (see discussion of exposure above), particularly when other sources of exposure are considered and when pesticides are applied directly to soil.[103]

Other studies have found impaired brood development[104] and increased rates of *Nosema* infection in honey bees exposed to sublethal pesticide levels.[105] Imidacloprid ingestion by stingless bee larvae at rates above 0.0056 μg/bee decreased survival rates, negatively affected development of a specific region of the bee brain called the mushroom body, and impaired walking behavior of newly emerged adult worker bees.[106] Scientists from Bayer Crop Science have criticized one of the studies linking neonicotinoid exposure with *Nosema* infection because it was conducted in the laboratory and not under field conditions.[107]

In another study, responsiveness to sucrose and waggle dancing (which communicates food locations to nest mates) was adversely affected when honey bees ingested a few nanograms of imidacloprid. The bees' preference for greater sucrose concentrations dissipated within an hour of ingestion.[108]

Finally, in another recent experiment, 16 hives of honey bees were fed corn syrup treated with imidacloprid. After 6 months, 15 of the 16 treated hives collapsed, while 4 untreated control hives remained healthy.[109]

A few recent studies have looked at bumble bees. One found that bumble bee colonies exposed in a laboratory to low levels of imidacloprid had a significantly reduced growth rate and an 85% reduction in queen production relative to untreated colonies. [110] Another study found that ingestion of low concentrations of neonicotinoids could adversely affect foraging behavior.[111] Bayer Crop Science argues that these results were obtained "under artificial conditions and are in conflict with" earlier studies.[112]

NEONICOTINOID REGISTRATION REVIEW AT EPA

Neonicotinoid pesticide registrations currently are being reviewed by EPA. EPA's website states:

> The neonicotinoids are a class of insecticides with a common mode of action that affects the central nervous system of insects, causing paralysis and death. All of the neonicotinoids were registered after 1984 and were not subject to reregistration. Some uncertainties have been identified since their initial registration regarding the potential environmental fate and effects of neonicotinoid pesticides, particularly as they relate to pollinators. Data suggest that neonicotinic residues can accumulate in pollen and nectar of treated plants and may represent a potential exposure to pollinators. Adverse effects data as well as beekill incidents have also been reported, highlighting the potential direct and/or indirect effects of neonicotinic pesticides. Therefore, among other refinements to ecological risk assessment during registration review, the Agency will consider potential effects of the neonicotinoids to honeybees and other pollinating insects.[113]

Review of clothianidin registrations began December 2011. EPA aims to review all neonicotinoids as a group.[114]

In the process of reviewing registrations for neonicotinoids, EPA is revising its risk assessment process. A letter from the Environmental Fate and Effects Division of the EPA Office of Chemical Safety and Pollution Prevention explains:

> The EPA is currently revising its process for assessing pesticide risks to bees to reflect advancements in the state of the science that underlie bee exposure and effects assessments. Interim guidance [EPA, 2011, Pesticides: Science and Policy, Interim Guidance on Honey Bee Data Requirements, http://www.epa.gov/pesticides /science/efed/policy_ guidance/team_authors /terrestrial_biology_tech_team/honeybee_ data_ interim_guidance.pdf] on factors to consider when evaluating exposure and effects to bees is available to ecological risk assessors. In 2012, the EPA will present to a FIFRA Scientific Advisory Panel (SAP) a proposed process for quantifying risks to honeybees and identifying exposure and effect studies needed to inform that process. Based on input from the SAP, the EPA will incorporate its revised assessment process to quantify risks to bees in a similar manner as that used to evaluate risks to other taxa.[115]

The draft risk assessment policy has been released to the public, and the materials distributed and discussed at the SAP meeting are available in the regulatory docket (EPA-HQ-OPP-2012-0543).

On March 20, 2012, EPA received a citizen petition asking it to suspend registrations for the insecticide clothianidin, among other demands.[116] The petition and numerous supplemental submissions of research reports and opinions are posted in the regulatory docket on the Internet at http://www.regulations.gov. EPA responded in part to the petition on July 17, 2012. EPA denied the request to suspend registrations "to prevent imminent harm" because the petitioners did not meet the burden of proof for registration suspension. However, EPA plans to revisit this decision and to evaluate the other demands in the near future as it reviews neonicotinoid registrations.

End Notes

[1] United Nations Environment Programme (UNEP), Global Honey Bee Colony Disorders and Other Threats to Insect Pollinators, UNEP Emerging Issues, 2010, http://www.unep.org /dewa/Portals/67/pdf/ Global_Bee_ Colony_Disorder_and_Threats_insect_pollinators.pdf.

[2] Ibid.

[3] For more information, see CRS Report RL33938, *Honey Bee Colony Collapse Disorder*.

[4] National Academy of Sciences, National Research Council, 2007, *Status of Pollinators in North America*, http://www.nap.edu/catalog.php?record_id=11761. Hereafter referred to as the 2007 NRC study.

[5] 2007 NRC study, p. 36; and S. Fecht, "Hive and Seek: Domestic Honeybees Keep Disappearing, but Are Their Wild Cousins in Trouble, Too?" *Scientific American*, May 8, 2012.

[6] Iowa State University, "Native Bees of North America," http://bugguide.net/node/view/475348.

[7] 2007 NRC study, p. 50; and Xerces Society, "Pollinator Conservation: Native Bee Biology," http://www. xerces.org/ native-bees/.

[8] Refers to bees that lay their eggs in another bee's nest; as their eggs hatch early, the cuckoo larvae eat the other bee's provisions.

[9] Staple crops (wheat, corn, and rice) do not rely on insect pollination and are mostly wind-pollinated.

[10] R. A. Morse and N. W. Calderone, *The Value of Honey Bees as Pollinators of U.S. Crops in 2000*, March 2000, Cornell University; and A.M. Klein, et. al., "Importance of pollinators in changing landscapes for world crops," *Proceedings of the Royal Society B: Biological Sciences*, Vol. 274, No. 1608, February 7, 2007.

[11] USDA, "Agroforestry: Sustaining Native Bee Habitat Crop Pollination," AF note-32, August 2006, http://plants.usda.gov/pollinators/Agroforestry_Sustaining_Native_Bee_Habitat_for_Crop_Pollination.pdf.

[12] R. A. Morse and N. W. Calderone, *The Value of Honey Bees as Pollinators of U.S. Crops in 2000*, March 2000, Cornell University, http://www.masterbeekeeper.org/pdf/pollination.pdf. This estimated value is measured according to the additional value of production attributable to honey bees, in terms of the value of the increased yield and quality achieved from honey bee pollination, including the indirect benefits of bee pollination required for seed production of some crops.

[13] R. Chaplin-Kramer, et al., "Value of Wildland Habitat for Supplying Pollination Services to Californian Agriculture," Rangelands, June 2011, Vol. 33(3): 33-41.

[14] National Honey Bboard, "Beekeepers and Honeybee Colonies," http://www.honey.com/nhb /media/press-kit/. Data from 2012 industry survey, *Bee Culture* magazine.

[15] A colony of bees is composed of between 250—50,000 individual bees (Source: D. Sammataro and A. Avitabile, *The Beekeepers Handbook*, 4th Edition), although most larger, healthy, managed hives will have between 20,000—50,000 bees.

[16] These operations are able to supply a large number of honey bee colonies during the critical phase of a crop's bloom cycle, when bees pollinate a crop as they fly from flower to flower collecting nectar and pollen, which they carry back to the nest.

[17] USDA, *2007 Census of Agriculture*, Table 31, http://www.agcensus.usda.gov/Publications /2007/Full_Report/ Volume_1,_Chapter_1_US/st99_1_029_031.pdf.

[18] Presentation to Congressional staff by Jeff Pettis, USDA/Agricultural Research Service, June 18, 2012.

[19] USDA, Honey, March 2012 http://usda.mannlib.cornell.edu/MannUsda/view DocumentInfo.do?document ID=1191.

[20] C. Souza, "What's the buzz about pollination?" *California Country* magazine, March/April 2011.

[21] See, for example, "Honey Bees and Blueberry Pollination," Fact Sheet 629, University of Maine Extension, April 2002.

[22] Penn State University, "A Year in the Life of a Migratory Honey Bee Colony;" M. Burgett, *1999 Pacific Northwest Honey Bee Pollination Survey*, Oregon State University; and Brenda Kiessling, presentation at Green Spring Garden Park's "EcoSavvy Symposium," Virginia, February 2012.

[23] See, for example, M. A. Aizen and L. D. Harder, "The Global Stock of Domesticated Honey Bees is growing Slower than Agricultural Demand for Pollination," *Current Biology*, May 2009.

[24] 2007 NRC study, p. 3.

[25] Ibid.

[26] J. Kim Kaplan, "Colony Collapse Disorder: An Incomplete Puzzle," *Agricultural Research* (USDA publication), July 2012, http://www.ars.usda.gov/is/AR/archive/jul12/colony 0712.htm. For other information, see CRS Report RL33938, *Honey Bee Colony Collapse Disorder*. The definition of CCD was recently revised to include other probably contributing factors, such as low levels of *Varroa* mite and other pathogens, such as *Nosema*, which appear to be associated with most winter colony losses. (Also see Benjamin Dinat, Jay D. Evans, Yan Ping Chen, et al., 2012, "Predictive markers of honey bee colony collapse," *PLoS ONE* , vol. 7, no. 2 (p. e32151).

[27] J. Kim Kaplan, op. cit.

[28] Ibid.

[29] Dennis vanEngelsdorp, Jay Hayes, and Jeff Pettis, 2009, "Preliminary Results: A Survey of Honey Bee Colonies Losses in the U.S. between September 2008 and April 2009." Survey based on about 20% of all U.S. colonies.

[30] Jeff Pettis, "Colony Collapse Disorder Affecting Honey Bee (*Apis mellifera*) Colonies," October 2008 presentation, http://www.epa.gov/pesticides/ppdc/2008/oct2008/session7-ccd.pdf. Survey based on 22 operations, managing 10% of U.S. colonies, and AIA surveys.

[31] D. vanEngelsdorp, J. Pettis, K. Rennich, et al., 2012, "Preliminary Results: Honey Bee Colony Losses in the U.S., Winter 2011-2012," http://beeinformed.org/2012/05/winter2012/; K. Rennich, J. Pettis, D. vanEngelsdorp, et al., 2012, "2011-2012 National Honey Bee Pests and Diseases Survey Report," http://www.aphis.usda. gov/plant_health/ plant_pest_info /honey_bees/downloads/2011_National_Survey_Report.pdf; and USDA, "Survey by USDA and Collaborators Reports Fewer Winter Honey Bee Losses," May 31, 2012, http://www.ars.usda.gov/is/pr/2012/ 120531.htm.

[32] See, for example, the University of California's backyard bee count (http://www.greatsunflower.org/), the University of Florida's "Native Buzz" project (http://www.ufnativebuzz.com/), and the "Bee Hunt" project organized by multiple university staff, federal agencies, and private partnerships (http://www.discoverlife.org /bee/index.html).

[33] 2007 NRC study, p. 7. See also 2007 NRC study, p. 88 (referring to *Bombus*, or bumble bees) ; p. 29 (referring to pollinators generally in North America); and p. 203 (referring to the absence of baseline data on pollinator status generally as an impediment to estimation of any decline).

[34] A well-known example of citizen-scientist-based data in North America is the annual Christmas bird count, which has been conducted for nearly a century in some areas. Data may not serve for some quantitative analyses, but may be much more reliable for the presence or absence of a bird, for example. In other studies, on arrival dates at bird feeders, or bees on spring flowers, the date on which a species is first observed could be another relatively reliable observation.

[35] I. Bartomeus, J.S. Ascher, D. Wagner, et al., 2011, "Climate-associated phenological advances in bee pollinators and bee-pollinated plants," *Proceedings of the National Academy of Sciences*, v. 108, n. 51, pp. 20645-20649. *Phenology* is the scientific study of biological phenomena that are cyclical or periodic. Examples include migration, hibernation, and nesting, as well as flowering times. Phenology is particularly important in analysis of climate change.

[36] The species studied were in four genera: *Andrena* (three species), *Bombus* (two species), *Colletes* (one species), and *Osmia* (four species).

[37] S.G. Potts, J.C. Biesmeijer, C. Kremen, et al., 2010, "Global pollinator declines: trends, impacts and drivers," *Trends in Ecology and Evolution*, v. 25, n. 6, pp. 345-353.

[38] Ibid.

[39] 2007 NRC report, p. 43.

[40] The University of Wyoming, through its Berry Center, sponsors the program. See http://www. uwyo.edu/berrycenter/ citizen-science/bumblebees.html for more information.

[41] S.A. Cameron, J.D. Lozier, and J.P. Strange, et al., 2011, "Patterns of widespread decline in North American bumble bees," *Proceedings of the National Academy of Sciences*, v. 108, n. 2, pp. 662-667.

[42] The authors also noted that some believe that *Nosema bombi* was introduced via commercially imported European bumble bees, and North American bumble bees may therefore have less resistance to the disease than European species. They compare the impacts of the pathogen to the fungal infections currently devastating wild populations of amphibians and bats.

[43] 2007 NRC report, p. 48.

[44] G. C. Venturieri, 2009, "The impact of forest exploitation on Amazonian stingless bees (Apidae Meliponini)," *Genetics and Molecular Research*, v. 8, n. 2, pp. 684-689.

[45] Ibid.

[46] H.V.V. Tomé, G.F. Martins, et al., 2012, "Imidacloprid-induced impairment of mushroom bodies and behavior of the native stingless bee Melipona quadrifasciata anthidioides," *PLoS ONE,* v. 7, n. 6, p. e38406, http://www.plosone.org/ article/info%3Adoi%2F10.1371%2 Fjournal.pone.0038406.

[47] Compiled by CRS from multiple sources (2007 NRC study; USDA, *Colony Collapse Disorder Progress Report*, CCD Steering Committee, June 2011; comments by Laurie Adams, North American Pollinator Protection Campaign (NAPPC) in a presentation to Congressional staff on June 18, 2012.

[48] A miticide is a pesticide intended to kill mites.

[49] USDA, *CCD Progress Report*, June 2011, http://www.ars.usda.gov/is/br/ccd/ccdprogress report2011.pdf.

[50] USDA, *CCD Progress Report*, June 2009, http://www.ars.usda.gov/is/br/ccd/ccdprogress report.pdf.

[51] D. vanEngelsdorp, J.D. Evans, C. Saegerman, et al., 2009, "Colony collapse disorder: A descriptive study," *PLOS One*, v. 4, n. 8, (August), e6481. doi:10.1371 /journal.pone. 0006481, http://www.plosone.org /article/ info:doi/10.1371/journal. pone.0006481.

[52] USDA, "Q&A: Colony Collapse Disorder," http://www.ars.usda.gov/News/docs.htm? docid=15572.

[53] Federal Insecticide, Fungicide, and Rodenticide Act, Section 2(u). 7 U.S.C. 136-136y. For more detailed information about pesticide laws and regulations, see CRS Report RL31921, *Pesticide Law: A Summary of the Statutes*.

[54] S. Kaiser, EPA, personal communication with CRS staff, December 16, 2011.

[55] Ingredients in pesticide products are categorized as active or inert. Active ingredients are those that are intended to control the pest, while inert ingredients, now generally known as "other ingredients," are used to deliver the active ingredients effectively to the pest. Other ingredients often are solvents or surfactants and often comprise the bulk of the pesticide product. Some inerts are known to be toxic, some are known to be harmless, and others have unknown toxicity.

[56] J.H. Cane and V.J. Tepedino, 2001, "Causes and extent of declines among native North American invertebrate pollinators: Detection, evidence, and consequences," *Conservation Ecology*, v. 5, n. 1, p. 1, http://www.ecologyandsociety.org/vol5/iss1/art1/.

[57] C.A. Johansen and D.F. Mayer, 1990, *Pollinator Protection, A Bee and Pesticide Handbook.*

[58] 2007 NRC study, p. 79.

[59] C.A. Mullin, M. Frazier, J.L. Frazier, et al., 2010, "High levels of miticides and agrochemicals in North American apiaries: Implications for honey bee health," *PLoS ONE*, v. 5, n. 3, p. e9754, http://www.plosone.org/article/info%3Adoi%2F10.1371%2Fjournal.pone.0009754.

[60] J. E. Cresswell, N. Desneux, and D. vanEngelsdorp, 2012, "Dietary traces of neonicotinoid pesticides as a cause of population declines in honey bees: an evaluation by Hill's epidemiological criteria," *Pest Management Science*, v. 68, pp. 819-827.

[61] L. Maxim and J.P. van der Sluis, 2010, "Expert explanations of honeybee losses in areas of extensive agriculture in France: Gaucho® compared with other supposed causal factors," *Environmental Research Letters,* v. 5, 014006, http://stacks.iop.org/ERL/5/014006.

[62] Ibid.

[63] Beebread is a pollen mixture stored in honeycomb cells and used with honey by bees as food. Beebread provides nutrition for the queen and brood.

[64] Trapped pollen refers to pollen collected at the entrance to the hive, removed from bees with the use of a trap.

[65] This is wax originating in the area of the hive where brood is reared.

[66] Foundation is the hexagonal sheets of wax and or wax/plastic combination that serves as a basis for comb construction by honey bees.

[67] C.A. Mullin, M. Frazier, J.L. Frazier, et al., 2010, "High levels of miticides and agrochemicals in North American apiaries: Implications for honey bee health," *PLoS ONE*, v. 5, n. 3, p. e9754.

[68] Ibid.

[69] J. Aufauvre, D.G. Biron, C. Vidau, et al. 2012, "Parasite-insecticide interactions: A case study of *Nosema ceranae* and fipronil synergy on honeybee," *Scientific Reports* 2, 326; C. Vidau, M, Diogon, J. Aufauvre, et al., 2011, "Exposure toe sublethal doses of fipronil and thiacloprid highly increases mortality of honeybees previously infected by *Nosem ceranae*," *PLoS* ONE, v. 6, n. 6, p. e21550; and C. Alaux, J. Brunet, C. Dussaubat, et al., 2012, "Interactions between *Nosema* microspores and a neonicotinoid weaken honeybees (*Apis mellifera*)," *Environmental Microbiology*, v. 12, n. 3. pp. 774-782.

[70] T. J. Ciarlo, C. A. Mullin, James L. Frazier, et al., 2012, "Learning impairment in honey bees caused by agricultural spray adjuvants," *PLoS ONE,* v. 7, n. 7, p. e40848.

[71] Ibid.

[72] D. vanEngelsdorp and M.D. Meixner, 2010, "A historical review of managed honey bee populations in Europe and the United States and factors that may affect them," *Journal of Invertebrate Pathology*, v. 103, Supplement 1, p. S80-S 95.

[73] L. Maxim and J.P. van der Sluis, 2010, "Expert explanations of honeybee losses in areas of extensive agriculture in France: Gaucho® compared with other supposed causal factors," *Environmental Research Letters,* v. 5, http://stacks.iop.org/ERL/5/014006.

[74] UNEP, *Global Honey Bee Colony Disorders and Other Threats to Insect Pollinators*, 2010.

[75] P. Jeschke, R. Nauen, M. Schindler, ,et al., 2011, "Overview of the status and global strategy for neonicotinoids," *Journal of Agricultural and Food Chemistry*, v. 59, pp. 2897-2908.

[76] C. Krupke, G. Hunt, B. Eitzer, et al., 2012, "Multiple routes of pesticide exposure for honey bees living near agricultural fields," *PLoS ONE*, v. 7, n. 1, p. e29268.

[77] Ibid.

[78] Ibid.

[79] G. C. Cutler and C. D. Scott-Dupree, 2007, "Exposure to clothianidin seed-treated canola has no long-term impact on honey bees," *Journal of Economic Entomology*, v. 100, pp. 765-772.

[80] J. M. Bonmatin, P. A. Marchand, R. Cahvet, et al., 2005, "Quantification of imidacloprid uptake in maize crops," *Journal of Agricultural and Food Chemistry*, v. 53, pp. 5336-5341.

[81] C. Krupke, G. Hunt, B. Eitzer, et al., 2012, "Multiple routes of pesticide exposure for honey bees living near agricultural fields," *PLoS ONE*, v. 7, n. 1, p. e29268.

[82] F. M. Laurent and E. Rathahao, 2003, "Distribution of [^{14}C] imidacloprid in sunflowers (*Helianthus annuus L.*) following seed treatment," *Journal of Agricultural and Food Chemistry*, v. 51, pp. 8005-8010.

[83] R. Schmuck, R. Schoning, A. Stork, et al., 2001, "Risk posed to honeybees (*Apis mellifera* L, Hymenoptera) by an imidacloprid seed dressing of sunflowers," *Pest Management Science*, v. 57, pp. 225-238.

[84] K. A. Stoner and B. D. Eitzer, 2012, "Movement of soil-applied imidacloprid and thiamethoxam into nectar and pollen of squash (*Cucurbita pepo*)," *PLoS ONE, v. 7, n. 6, p. e39114.*

[85] G. P. Dively and A. Kamel, 2012, "Insecticide residues in pollen and nectar of a cucurbit crop and their potential exposure to pollinators," *Journal of Agricultural and Food Chemistry*, v. 60, pp. 4449-4456.

[86] K. A. Stoner and B. D. Eitzer, 2012, "Movement of soil-applied imidacloprid and thiamethoxam into nectar and pollen of squash (*Cucurbita pepo*)," *PLoS ONE, v. 7, n. 6, p. e39114.*

[87] For example, see H.V.V. Tomé, G.F. Martins, M. August, et al., 2012, "Imidacloprid-induced impairment of mushroom bodies and behavior of the native stingless bee *Melipona quadrifasciata anthidioides*," *PLoS ONE*, v. 7, n. 6, p. e38406; D.M. Eiri and J. C. Nieh, 2012, "A nicotinic cacetylcholine receptor agonist affects honey bee sucrose responsiveness and decreases waggle dancing," *The Journal of Experimental Biology*, v. 215, n. 12, pp. 2022-2029; A. Decourtye, E. Lacassie, and M. Pham-Delegue, 2003, "Learning performances of honey bees (*Apis mellifera L.*) are differentially affected by imidacloprid according to the season." *Pest Management Science*, v. 59, pp. 269-278; or V. Mommaerts, S. Reynders, J. Boulet, et al., 2010, "Risk assessment for side-effects of neonicotinoids against bumblebees with and without impairing foraging behavior," *Ecotoxicology*, v. 19, pp. 207-215.

[88] Chensheng Lu, K.M. Warchol, and R.A. Callahan, 2012, "In situ replication of honey bee colony collapse disorder," *Bulletin of Insectology*, v. 65, n. 1 99-106.

[89] G. DeGrandi-Hoffman, D. Sammataro, and R. Simonds, 2012, "Are agrochemicals present in high fructose corn syrup fed to honey bees (*Apis mellifera* L.)?" *Jurnal of Apicultural Research*, v. 51, n. 4, pp. 371-372.

[90] D. Laurino, M. Porporato, A. Patetta, et al., 2011, "Toxicity of neonicotinoid insecticides to honey bees: laboratory tests," *Bulletin of Insectology*, v. 64, n. 1, pp. 107-113.

[91] S. P. Bradbury, Office of Pesticide Programs, Letter to Peter T. Jenkins, Center for Food Safety and International Center for Technology Assessment, July 17, 2012. Hereafter referred to as Bradbury response to citizen petition.

[92] J. DeCant and M. Barrett, 2010, "Environmental Fate and Ecological Risk Assessment for the Registration of CLOTHIANIDIN for Use as a Seed Treatment on Mustard Seed (Oilseed and Condiment) and Cotton," EPA/Office of Prevention, Pesticides and Toxic Substances, Office of Pesticide Programs, Washington, DC, p. 14.

[93] Ibid.

[94] A. Tapparo, D. Marton, C. Giorio, et al., 2012, "Assessment of the environmental exposure of honeybees to particulate matter containing neonicotinoid insecticides coming from corn coated seeds," *Environmental Science & Technology*, v.46, pp. 2592-2599.

[95] V. Girolami, M. Marzaro, L. Vivian, et al., 2012, "Fatal powdering of bees in flight with particulates of neonicotinoids seed coating and humidity implication," *Journal of Applied Entomology*, v. 136, n. 1-2, pp. 17-26.

[96] D. Cox-Foster and D. vanEngelsdorp, "Solving the Mystery of the Vanishing Bees," *Scientific American*, March 31, 2009; Northwest Coalition for Alternatives to Pesticides, "Imidacloprid, Fact Sheet," *Journal of Pesticide Reform*, Spring 2001, http://www.pesticide.org/imidacloprid.pdf; and Apiculteurs de France, "Composite Document of Present Position Relating to Gaucho, Sunflower and Bees," http://www.beekeeping.com/articles/us/gaucho/ manifestation_paris_us.htm.

[97] Joe Cummins, "Neoniccotinoid insecticides used in seed dressing may be responsible for the collapse of honeybee colonies," April 24, 2007, http://www.organicconsumers.org /articles/article_4972.cfm.

[98] BNA's *International Environmental Law Committee Newsletter*, vol. 11, no. 1, February 2009, http://www.abanet.org/environ/committees/intenviron/newsletter/feb09/IELC_Feb09.pdf.

[99] M. Henry, M. Beguin, F. Requier, et al., 2012, "A common pesticide decreases foraging success and survival in honey bees," *Science*, v. 336, n. 6079, pp. 348-350.

[100] Bayer Crop Science, "Overview of recent publications on neonicotinoids and pollinators." Paper received personally from Jean Reimers (CropLife America), June 20, 2012.

[101] C.W. Schneider, J. Tautz, B. Grünewald, et al., 2012, "RFID Tracking of sublethal effects of two neonicotinoid insecticides on the foraging behavior of *Apis mellifera*." PLoS ONE, v. 7, n. 1, p. e30023, http://www.plosone.org/ article/info%3Adoi%2F10.1371% 2Fjournal.pone. 0030023.

[102] Ibid.

[103] K. A. Stoner and B. D. Eitzer, 2012, "Movement of soil-applied imidacloprid and thiamethoxam into nectar and pollen of squash (*Cucurbita pepo*)," *PLoS ONE, v. 7, n. 6, p. e39114.*

[104] Judy Y. Wu, Carol M. Anelli, and Walter S. Sheppard, 2011, "Sub-lethal effects of pesticide residues in brood comb on worker honey bee (*Apis mellifera*) development and longevity," *PLoS ONE* v., 6, n. 2, p. e14720.

[105] J.Y. Wu, M.D. Smart, C.M. Anelli, et al., 2012, "Honey bees (*Apis mellifera*) reared in brood combs containing high levels of pesticide residues exhibit increased susceptibility to *Nosema* (Microsporidia) infection," *Journal*

of Invertebrate Pathology, v. 109, pp. 326-329; and J. Pettis, D. vanEngelsdorp, J. Johnson, et al., 2012, "Pesticide exposure in honey bees results in increased levels of the gut pathogen *Nosema*," *Naturwissenschaften*, v. 99, pp. 153- 158.

[106] H. V. V. Tomé, G. F. Martins, M. August, et al., 2012, "Imidacloprid-induced impairment of mushroom bodies and behavior of the native stingless bee Melipona quadrifasciata anthidioides," PLoS ONE, v. 7, n. 6, p. e38406.

[107] Bayer Crop Science, "Overview of recent publications on neonicotinoids and pollinators." Paper received personally from Jean Reimers (CropLife America), June 20, 2012.

[108] D. M. Eiri and J. C. Nieh, 2012, "A nicotinic cacetylcholine receptor agonist affects honey bee sucrose responsiveness and decreases waggle dancing," *The Journal of Experimental Biology*, v. 215, n. 12, pp. 2022-2029.

[109] Chensheng Lu, K.M. Warchol, and R.A. Callahan, 2012, "In situ replication of honey bee colony collapse disorder," *Bulletin of Insectology*, v. 65, n. 1 99-106.

[110] P.R. Whitehorn, S. O'Connor, D. Goulson, et al. 2012, "Neonicotinoid pesticide reduces bumble bee colony growth and queen production," *Science*, v. 336, p. 351-352.

[111] V. Mommaerts, S. Reynders, J. Boulet, et al., 2010, "Risk assessment for side-effects of neonicotinoids against bumblebees with and without impairing foraging behavior," *Ecotoxicology*, v. 19, pp. 207-215.

[112] Bayer Crop Science, "Overview of recent publications on neonicotinoids and pollinators." Paper received personally from Jean Reimers (CropLife America), June 20, 2012.

[113] EPA, Pesticides, Registration Review, Program Highlights, Groups of Related Pesticides Beginning Registration Review, Neonicotinoids, May 9, 2012, http://www.epa.gov /oppsrrd1/registration_review/highlights.htm, Dec. 4, 2012.

[114] The status of the review and relevant documents is available at http://www.regulations.gov (docket: EPA-HQ-OPP2011-0865). EPA review of imidacloprid registrations began December 2008 (docket: EPA-HQ-OPP-2008-0844). The schedule for other reviews is at http://www.epa.gov/oppfead1/cb/ppdc/2012/may/session-9-reg-review-update.pdf.

[115] EPA, Environmental Fate and Effects Division (EFED), EFED Response to Comments Submitted to the Clothianidin Registration Review Docket (Docket ID: EPA-HQ-OPP-2011-0865), June 11, 2012, p. 3.

[116] S.P. Bradbury, EPA Office of Pesticide Programs, Letter to Peter T. Jenkins, Center for Food Safety and International Center for Technology Assessment, "Clothianidin Emergency Citizen Petition dated March 20, 2012," July 17, 2012, http://www.epa.gov/opp00001 /about/intheworks/epa-respns-to-clothianidin-petition-17july12.pdf.

In: Bee Health: Factors, Analyses, and Research Progress ISBN: 978-1-62618-044-4
Editor: Roger Dunham

Chapter 2

HONEY BEE COLONY COLLAPSE DISORDER*

Renée Johnson

SUMMARY

Starting in late 2006, commercial migratory beekeepers along the East Coast of the United States began reporting sharp declines in their honey bee colonies. Because of the severity and unusual circumstances of these colony declines, scientists named this phenomenon colony collapse disorder (CCD). Reports indicate that beekeepers in most states have been affected. Overall, the number of managed honey bee colonies dropped an estimated 35.8% and 31.8% in the winters of 2007/2008 and 2006/2007, respectively. Preliminary loss estimates for the 2008/2009 winter are reported at 28.6%. To date, the precise reasons for colony losses are not yet known.

Honey bees are the most economically valuable pollinators of agricultural crops worldwide. Scientists at universities and the U.S. Department of Agriculture (USDA) frequently assert that bee pollination is involved in about one-third of the U.S. diet, and contributes to the production of a wide range of fruits, vegetables, tree nuts, forage crops, some field crops, and other specialty crops. The monetary value of honey bees as commercial pollinators in the United States is estimated at about \$15-\$20 billion annually.

Honey bee colony losses are not uncommon. However, losses in recent years differ from past situations in that colony losses are occurring mostly because bees are failing to return to the hive (which is largely uncharacteristic of bee behavior); bee colony losses have been rapid; colony losses are occurring in large numbers; and the reason(s) for these losses remains largely unknown.

Based on the available research over the past few years on the numerous possible causes of CCD, USDA concluded in its 2007-2008 progress report (released in June 2009) that "it now seems clear that no single factor alone is responsible for the malady." This has led researchers to further examine the hypothesis that CCD may be "a syndrome caused by many different factors, working in combination or synergistically." Currently, USDA states, researchers are focusing on three major possibilities:

* This is an edited, reformatted and augmented version of Congressional Research Service, Publication No. RL33938, dated January 7, 2010.

- pesticides that may be having unexpected negative effects on honey bees;
- a new parasite or pathogen that may be attacking honey bees, such as the parasite *Nosema ceranae* or viruses; and
- a combination of existing stresses that may compromise the immune system of bees and disrupt their social system, making colonies more susceptible to disease and collapse. Stresses could include high levels of infection by the *Varroa* mite; poor nutrition due to apiary overcrowding, pollination of crops with low nutritional value, or pollen or nectar scarcity; exposure to limited or contaminated water supplies; and migratory stress.

Funding for honey bee and CCD research at USDA's Agricultural Research Service (ARS) has increased sharply, following both the enactment of the 2008 farm bill (P.L. 110-246) and the FY2009 and FY2010 appropriations process (P.L. 111-8 and P.L. 111-80, respectively). These legislative actions contained additional provisions that would, among other things, provide additional funding for research and conservation programs addressing honey bees and pollinators. Total ARS funding for honey bee and CCD research averaged more than $7.7 million each in FY2007 and FY2008, increasing to $8.3 million in FY2009 and $9.8 million for FY2010.

INTRODUCTION

Starting in 2006, commercial migratory beekeepers along the East Coast of the United States began reporting sharp declines in their honey bee colonies. Because of the severity and unusual circumstances of these colony declines, scientists named this phenomenon colony collapse disorder (CCD). Current reports indicate that beekeepers in most states have been affected. Overall, the number of managed honey bee colonies dropped an estimated 35.8% and 31.8% in the winters of 2007/2008 and 2006/2007, respectively.[1] Preliminary loss estimates for the 2008/2009 winter are reported at 28.6%. To date, the precise reasons for colony losses are not yet known.

This report provides an overview of the importance of honey bee pollination to U.S. agricultural production, especially specialty crops. It describes the extent and symptoms of CCD and how it differs from previous honey bee colony losses, describing some of the reasons why scientists believe honey bee colonies are being affected by CCD. Finally, the report discusses policy options and actions that Congress has taken to address this issue.

IMPORTANCE OF HONEY BEE POLLINATION

Honey bees (*Apis mellifera*) are the most economically valuable pollinators of agricultural crops worldwide and are the only bee species kept commercially in the United States.[2] In the United States, bee pollination of agricultural crops is said to account for about one-third of the U.S. diet, and to contribute to the production of a wide range of high-value fruits, vegetables, tree nuts, forage crops, some field crops, and other specialty crops.[3]

The monetary value of honey bees as commercial pollinators in the United States is estimated at about $15 billion annually[4] (Table 1). Some studies report the estimated value of honey bee pollination at as much as $20 billion annually.

This estimated value is measured according to the additional value of production attributable to honey bees, in terms of the value of the increased yield and quality achieved from honey bee pollination, including the indirect benefits of bee pollination required for seed production of some crops. About one-third of the estimated value of commercial honey bee pollination is in alfalfa production, mostly for alfalfa hay. Another nearly 10% of the value of honey bee pollination is for apples, followed by 6%-7% of the value each for almonds, citrus, cotton, and soybeans.

Overall, pollinator-dependent crops are reported to make up an estimated 23% of total U.S. agricultural production in 2006, up from an estimated 14% in the 1960s.[5]

Table 1. Estimated Value of the Honey Bee to U.S. Crop Production, 2000 Estimates

Crop Category (ranked by share of honey bee pollinator value)	Dependence on Insect Pollination	Proportion of Pollinators That Are Honey Bees	Value Attributed to Honey Bees[a] ($ millions)	Major Producing States[b]
Alfalfa, hay & seed	100%	60%	4,654.2	CA, SD, ID, WI
Apples	100%	90%	1,352.3	WA, NY, MI, PA
Almonds	100%	100%	959.2	CA
Citrus	20%-80%	10%-90%	834.1	CA, FL, AZ, TX
Cotton (lint & seed)	20%	80%	857.7	TX, AR, GA, MS
Soybeans	10%	50%	824.5	IA, IL, MN, IN
Onions	100%	90%	661.7	TX, GA, CA, AZ
Broccoli	100%	90%	435.4	CA
Carrots	100%	90%	420.7	CA, TX
Sunflower	100%	90%	409.9	ND, SD
Cantaloupe/honeydew	80%	90%	350.9	CA, WI, MN, WA
Other fruits & nuts[c]	10%-90%	10%-90%	1,633.4	—
Other vegetables/melons[d]	70%-100%	10%-90%	1,099.2	—
Other field crops[e]	10%-100%	20%-90%	70.4	—
Total	—	—	14,564	—

Source: Compiled by CRS using values reported in R. A. Morse, and N.W. Calderone, *The Value of Honey Bees as Pollinators of U.S. Crops in 2000*, March 2000, Cornell University, http://www.masterbeekeeper.org/pdf/ pollination.pdf.

a Attributed value is the additional value of production attributable to honey bees, in terms increased yield and quality achieved from honey bee pollination, including the indirect benefits of bee pollination required for seed production of some crops. Calculated from total average production value (1996-1998).

b For most commodities, major producing states reflect reported 2006 production (http://www.nass.usda.gov/ QuickStats/). Melon production is based on reported 2002 harvested acreage.

c Apricots, avocados, blueberries, brambleberries, cherries, cranberries, grapes, kiwi fruit, macadamia nuts, olives, peaches, pears, nectarines, plums, and strawberries.

d Asparagus, cauliflower, celery, cucumbers, pumpkins, squash, watermelon, and vegetable seeds.

e Peanuts, canola (rapeseed), and sugarbeets.

A number of agricultural crops are almost totally (90%-100%) dependent on honey bee pollination, including almonds, apples, avocados, blueberries, cranberries, cherries, kiwi fruit, macadamia nuts, asparagus, broccoli, carrots, cauliflower, celery, cucumbers, onions, legume seeds, pumpkins, squash, and sunflowers. Other specialty crops also rely on honey bee pollination, but to a lesser degree. These crops include apricot, citrus (oranges, lemons, limes, grapefruit, tangerines, etc.), peaches, pears, nectarines, plums, grapes, brambleberries, strawberries, olives, melon (cantaloupe, watermelon, and honeydew), peanuts, cotton, soybeans, and sugarbeets.[6]

In the United States, most pollination services are provided by commercial migratory beekeepers who travel from state to state and provide pollination services to crop producers. These operations are able to supply a large number of bee colonies during the critical phase of a crop's bloom cycle, when honey bees pollinate a crop as they fly from flower to flower collecting nectar and pollen, which they carry back to the nest.[7] Data from the U.S. Department of Agriculture's (USDA) *2007 Census* indicate that there were about 28,000 operations with 2.9 million bee colonies in the United States.[8] The majority of these, more than 2 million bee colonies, are reported to belong to commercial migratory beekeepers. The Dakotas accounted for a combined 27% of all bee colonies. Another one-fifth of all colonies are in California (about 14%) and Florida (7%).[9] Montana, Minnesota, Idaho, and Texas accounted for 4%-5% each of all colonies nationwide. Other states with a large number of bee colonies were Michigan, Oregon, Georgia, Nebraska, New York, Washington, Wisconsin, and Wyoming, with about 2% each. Although these operations also produce honey for commercial sale, it is their value as crop pollinators that provides the greatest economic impact in the production of food and feed crops.

It is estimated that, each year, more than 2 million bee colonies are rented for U.S. crop pollination. Available limited information indicates that the greatest number of honey bee colony rentals are for apple and almond production, followed by clover seed, cherries, and pears.[10] About one-half of the nation's honey bee colonies (an estimated 1.3 million colonies) are used to pollinate California's current 550,000 acres of almond trees, and this need is projected to grow to 1.5 million colonies by 2010.[11] Both locally and globally, there are concerns that growth in the availability of honey bee stocks is not keeping pace with growing agricultural demands for pollination services.[12]

Rental fees collected by commercial beekeepers for pollination services may vary by crop type, and often tend to be lower for some seed crops and higher for berry and tree crops. In recent years, pollination fees paid by crop producers have increased. For example, fees paid by California's almond industry have risen from a reported $35 per colony in the late 1990s to about $75 per colony in 2005.[13] More recent estimates of fees for pollinating almond trees are even higher, at $150 per colony or more. Among the reasons for higher pollination fees are expanding almond acreage and relatively high honey prices, but also fewer available honey bees for pollination due, in part, to colony declines and bee mortalities.

Extent and Symptoms of Colony Collapse Disorder

Starting in the last three months of 2006, a seemingly new phenomenon began to occur based on reports of an "alarming" number of bee colony losses and die-off along the East Coast. By the end of 2006, beekeepers on the West Coast also began to report "unprecedented" losses.[14] Available estimates indicate that beekeepers in 35 states have been affected (Figure 1). Because of the severity and lack of precedent, scientists coined a new term, Colony Collapse Disorder (CCD), for this phenomenon.

Source: Bee Alert Inc., "Map of U.S. States Reporting Colony Collapse Disorder," http://beealert.blackfoot.net/ ~beealert/USshaded.pdf. Shaded areas show reported affected states. This is the most recent update (as of December 2009).

Figure 1. Colony Collapse Disorder, Affected States.

Much of the current research on CCD is being conducted by scientists at USDA's Agriculture Research Service (Beltsville bee laboratory), Pennsylvania State University, the University of Montana, and the Pennsylvania and Florida Departments of Agriculture, among others. Many of these researchers also participate in the CCD Working Group, which includes Bee Alert Inc., the Florida and Pennsylvania Departments of Agriculture, Pennsylvania State University, and USDA. Up-to-date information is regularly posted to the website of the Mid-Atlantic Apiculture Research and Extension Consortium (MAAREC), which represents beekeeping associations in New Jersey, Maryland, Delaware, Pennsylvania, and West Virginia.

Past Honey Bee Population Losses

Honey bee colony losses are not uncommon. A recent report by the National Research Council (NRC) documents the extensive literature on honey bee population losses due to bee pests, parasites, pathogens, and disease. Most notable are declines due to two parasitic mites, the so-called vampire mite (*Varroa destructor*) and the tracheal mite (*Acarapis woodi*), and also colony declines due to the pathogen *Paenibacillus larvae*.[15] Other reasons for bee colony declines reported by the NRC include interspecific competition between native and introduced bees, pathogen spillover effects, habitat loss, invasive plant species that reduce nectar- and pollen-producing vegetation, bee genetics, and pesticides, among other factors.

Mite infestations are a relatively new occurrence. The 1980s saw two periods of large die-offs due to *Varroa* and tracheal mites: The first *Varroa* mite infestation was reported in 1987; tracheal mites were first detected in 1984.[16] *Varroa* mites are also said to have eliminated most feral bee colonies in the mid-1990s.[17] *Varroa* parasitism affects both worker bees and male larvae and can affect the ability of the queen to reproduce. It is associated with viral pathogens and if left untreated can cause colony mortalities usually within six months to two years after the initial infestation. Less is known about the effects of the tracheal mite. The pathogen *Paenibacillus larvae* is the most serious honey bee pathogen and causes American foulbrood (AFB), which is a disease of larval honey bees. AFB resulted in large colony losses in the 1940s, but its incidence has been reduced by the use of antibiotics and increased apiary inspection programs. Nevertheless, mite and pathogen infestations have likely raised beekeeper operating costs to pay for miticides and/or antibiotics, labor and expenses for treatment, improved management and inspection, and colony replacement of dead bees.

Symptoms similar to those observed for CCD have been described in the past, and heavy losses have been documented. It is still not clear whether the current colony losses are being caused by the same factors or if new contributing factors are involved.[18] MAAREC also reports that large beekeeper operations may have experienced higher than normal losses compared with the past few years, and heavy overwintering losses were reported in 2003-2004 for many northern beekeepers.

Overall, USDA reports that bee colony losses have averaged 17%-20% per year since the 1990s, attributable to a variety of factors, such as mites, diseases, and management stress. By comparison, bee colony losses between the winters of 2006/2007 and 2007/2008 averaged more than 30% during the year.[19]

Current Colony Losses from Available Surveys

The first report of CCD was in mid-November 2006 by a Pennsylvania beekeeper overwintering in Florida. By February 2007, large commercial migratory beekeepers in several states had reported heavy losses associated with CCD. Their reports of losses varied widely, ranging from 30% to 90% of their bee colonies; in some cases beekeepers feared loss of nearly all of their colonies.[20] Surviving colonies were reportedly weakened and might no longer be viable to pollinate or produce honey. Losses were reported in migratory operations wintering in California, Florida, Oklahoma and Texas. In late February, some larger non-migratory beekeepers in the mid-Atlantic and Pacific Northeast regions also reported significant losses of more than 50%.[21] Bee colony losses also were reported in five Canadian provinces, several European countries, and countries in South and Central America and Asia.

In March 2007, the Apiary Inspectors of America (AIA) conducted a survey of its members in 15 states.[22] The survey tracked changes from September 2006 and March 2007. Overall, responding beekeepers suffered an average loss of 38% of their colonies during the winter of 2006-2007. If these losses are representative of the nation, between 651,000 and 875,000 of the nation's estimated 2.4 million colonies were lost over that winter.[23] While a majority of losses were attributable to known causes, approximately 25% of beekeepers are believed to have suffered from CCD.[24] The survey indicated that, among the beekeepers surveyed, more than 50% reported "abnormally heavy losses" with total colony losses of 55%. This compared to those reporting "normal losses" with total colony losses of 16%. Of the responding beekeepers, about one-fourth reported conditions associated with CCD.[25] Beekeeping operations experiencing CCD-like conditions reported losses of 45% of their managed bee colonies. Among the leading causes reported by most affected commercial beekeeping operations were pest diseases.

A 2007 survey conducted by Bee Alert Technology, Inc., showed that, among the beekeepers surveyed, more than 40% reported "severe losses," with losses of nearly 60% of their colonies.[26] Another 48% reported average or lower losses. Smaller operations with less than 100 colonies are more likely to have suffered more severe losses than normal. Respondents were also asked to indicate whether the general cause for colony losses was due to overwinter losses, mites, pesticide exposure, or colony disappearance (or CCD). Among all respondents, colony losses due to disappearance (43%) and overwintering (37%) accounted for the greatest share of total losses, with mites and disease accounting for another 15%. Pesticides were indicated to account for a relatively small share (4%) of surveyed colony failures, regardless of operation size. This compares to other previous estimates of winter losses from various different surveys showing overall colony losses of about 30% during the period 2000-2006, mostly associated with losses due to *Varroa* mites.[27]

Survey information reported by USDA and AIA indicate that the number of managed honey bee colonies dropped an estimated 35.8% in the 2007/2008 winter and 31.8% in the 2006/2007 winter.[28] Preliminary loss estimates for the 2008/2009 winter are reported at 28.6%. This survey data also indicates that 15% of all the colonies lost during the 2008/2009 winter died with symptoms of CCD, compared to a 60% colony loss with CCD-like symptoms in the winter of 2007/2008. Although more recent estimates reflect a possible decrease in the rate of managed colony losses, USDA asserts that this rate of loss remains unsustainable.[29] Other information from USDA indicates that colony losses range widely depending on area, from 7% to 80% loss at some surveyed operations.[30]

How CCD Differs from Past Bee Colony Losses

Current bee colony losses seem to differ from past losses in that colony losses are occurring mostly because bees are failing to return to the hive (which is largely uncharacteristic of bee behavior); bee colony losses have been rapid; colony losses are occurring in large numbers; and the reason why these losses are occurring remains still largely unknown. The phenomenon was first called "Fall-Dwindle Disease," but was renamed because of the unusual characteristics of the colony declines. Moreover, the condition is not only seasonal but manifests itself throughout the year. The term "dwindle" implies a gradual loss, whereas CCD onset is sudden. Also, the term "disappearance" is used to describe other types of conditions, which differ from the symptoms currently being associated with CCD. Finally, the term "disease" is usually associated with a biological agent, but none has yet been identified.[31]

Symptoms of Colony Collapse Disorder

The symptoms of CCD, based on the available research, include the following: [32]

- rapid loss of adult worker bees,
- few or no dead bees found in the hive,

- presence of immature bees (brood),
- small cluster of bees with live queen present, and
- pollen and honey stores in hive.

Among the key symptoms of CCD in collapsed colonies is that the adult population is suddenly gone without any accumulation of dead bees. The bees are not returning to the hive but are leaving behind their brood (young bees), their queen, and maybe a small cluster of adults. What is uncharacteristic about this situation is that the honey bee is a very social insect and colony-oriented, with a complex and organized nesting colony. Failing to return to the hive is considered highly unusual. An absence of a large number of dead bees makes an analysis of the causes of CCD difficult.

Also there is little evidence that the hive may have been attacked. In actively collapsing colonies, an insufficient number of adult bees remain to care for the brood. The remaining workforce seems to be made up of young adult bees.

The queen is present, appears healthy and is usually still laying eggs, but the remaining cluster is reluctant to consume feed provided by the beekeeper, and foraging is greatly reduced.

Possible Causes of Colony Collapse Disorder

The initial scientific search of the possible factors involved in CCD focused on four areas:[33]

- pathogens,
- parasites,
- environmental stresses, and
- bee management stresses such as poor nutrition.

Early on, researchers had tentatively removed some practices and conditions from the list of possible causes of CCD. These included feeding practices, chemicals used by beekeepers (such as antibiotics and miticides), use of bees (primarily for honey production versus pollination), and queen source.[34] However, the scientists researching this phenomenon note that these could contribute to the risk of bee colonies developing CCD. Some scientists also wonder whether a combination of the stressors, including mites, disease, and nutritional stress, are interacting to weaken bee colonies and are allowing stress-related pathogens, such as fungi, thus causing a final collapse.[35] Others note the possible role of miticide resistance in bees.

Others have speculated that because most of the reported colony losses are among large commercial migratory operations, which may move bees two to five times during a growing season, the current disorder may be the result of accumulated stress, and factors such as confinement and temperature fluctuations. These stresses may increase the colony's susceptibility to disease and may also increase its potential exposure to other diseases and parasites.[36] A 10% die-off is not uncommon following transportation, with losses of 30% possible. As outlined in USDA's 2007-2008 progress report, the available research over the past few years on the numerous possible causes for CCD has led USDA and university researchers to conclude that "no single factor alone is responsible" for CCD.[37] This has led researchers to further examine the hypothesis that CCD may be "a syndrome caused by many different factors, working in combination or synergistically," including "an interaction between pathogens and other stress factors." [38] Currently, USDA states, researchers are focusing on three major possibilities:[39]

1. pesticides that may be having unexpected negative effects on honey bees;
2. a new parasite or pathogen that may be attacking honey bees, such as the parasite *Nosema ceranae* or viruses; and
3. a combination of existing stresses that may compromise the immune system of bees and disrupt their social system, making colonies more susceptible to disease and collapse. Stresses could include high levels of infection by the *Varroa* mite; poor nutrition due to apiary overcrowding, pollination of crops with low nutritional value, or pollen or nectar scarcity; exposure to limited or contaminated water supplies; and migratory stress.

Parasites, Mites, and Disease

Initially, the potential causes of CCD, as reported by the scientists researching this phenomenon, were thought to include but not be limited to parasites, mites, and disease loads in the bees and brood; emergence of new or newly more virulent pathogens, such as fungal diseases; poor nutrition among adult bees; lack of genetic diversity and lineage of bees; level of stress in adult bees, as indicated by stress-induced proteins (e.g., transportation and confinement of bees, overcrowding, or other environmental or biological stressors); chemical residue/contamination in the wax, food stores, and/or bees, including acute or cumulative exposure to new types of agricultural pesticides as well as exposure to chemicals that beekeepers use to control mites; and a combination of these and/or other factors.[40]

In July 2007, USDA reported that theories about the causes of CCD were focused on increased losses due to the *Varroa* mite; new or emerging diseases, especially mortality by a new species of a single-celled parasite *Nosema ceranae*; pesticide exposure; and potential immune-suppressing stress on bees due to one or a combination of these factors.[41] In September 2007, a research team that included USDA published the results of a genetic screening of CCD-affected honey bee colonies and non-CCD-affected hives.[42] The only pathogen found in nearly all samples (96.1%) from CCD-affected colonies, but not in non-CCD colonies, was the Israeli acute paralysis virus (IAPV), a dicistrovirus that can be transmitted by the *Varroa* mite.[43] USDA considers this research to have identified IAPV as a marker of CCD, since it is found in affected bees, but not to have identified IAPV as the cause of CCD; however, this research indicates there is a strong correlation of the appearance of IAPV and CCD together.[44]

High levels of bacteria, viruses, and fungi have been found in the guts of the recoverable dead bees. Early evidence does suggest the possible presence of a pathogen, given that some bee colonies have recovered once their bee boxes were irradiated.[45] Researchers have found the fungus *Nosema ceranae* and other pathogens such as chalkbrood in some affected hives throughout the country.[46] Some researchers have speculated that these high infection levels may be compromising the immune system of the honey bees, resulting in immune deficiencies in bees that may be among the possible causes for bee mortalities and disappearance.[47] A 2009 study by researchers in Spain found further evidence that infection in bees by *Nosema ceranae* may be among the primary causes of CCD.[48]

Pesticides and Chemicals

Of the possible causes of CCD being examined, one that has become the subject of debate is whether certain chemicals or combinations of chemicals could be contributing to CCD, including some pesticides and possibly some fungicides. Scientists have long been concerned that pesticides may have sub-lethal effects on bees, not killing them outright but instead impairing their development and behavior.[49]

One class of insecticide being studied are neonicotinoids, which contain the active ingredient imidacloprid, and similar other chemicals, such as clothianidin and thiamethoxam. Honey bees are thought possibly to be affected by such chemicals, which are known to work their way through the plant up into the flowers and leave residues in the nectar and pollen (which is the food for young, developing bees). The scientists studying CCD have tested samples of pollen and have indicated findings of a broad range of substances, including insecticides, fungicides, and herbicides.[50] These scientists note that the doses taken up by bees are not lethal, but they are concerned about possible chronic problems caused by long-term exposure. As noted by the NRC, some studies report sublethal effects of pesticides on bee foraging behavior that may impair the navigational and foraging abilities of honey bees.[51]

Concerns about imidacloprid, as reported by beekeeping associations in the United Kingdom and France[52] and by some U.S. beekeepers,[53] have focused on its potential to affect complex behaviors in insects, including flight, navigation, olfactory memory, recruitment, foraging, and coordination. However, the NRC and some scientists who study CCD note there is conflicting information about the effect of these pesticides on honey bees. Still, the U.S. Environmental Protection Agency has identified some of these chemicals as highly toxic to honey bees,[54] and use of some of these pesticides has reportedly been discontinued in parts of Europe because of their potential effects on pollinators.[55] However, bee colony losses are also occurring in Europe, where these chemicals are reportedly no longer used. In the United States, the Organic Consumers Association reports that bee colony losses are not occurring at organic beekeeping operations.[56]

Nevertheless, a number of environmental groups are taking legal action to highlight the possibility that pesticides and chemical loads may be contributing to colony declines. For example, in December 2009, a federal court in New York invalidated the U.S. Environmental Protection Agency's (EPA's) approval of the pesticide spirotetramat and ordered the agency to reevaluate the chemical, as a result of a suit filed by the Natural Resources Defense Council (NRDC) and the Xerces Society.[57] The pesticide, manufactured by Bayer CropScience under the trade names Movento and Ultor, is thought to be potentially harmful to honey bees. NRDC also filed a lawsuit against the EPA in August 2008 to obtain information that they allege the U.S. government is withholding about the risks posed by pesticides to honey bees. NRDC claims that EPA has evidence of connections between pesticides and CCD.[58] Also in August 2008, a German coalition group brought legal charges against Bayer AG, accusing them of "marketing dangerous pesticides" and contributing to bee colony declines.[59] The coalition filed the charge in cooperation with German beekeepers who claim they lost hives because of the Bayer pesticide clothianidin dating back to May 2008. Some countries, including Germany, Italy and France, reportedly are either considering or have already instituted full or partial bans of neonicotinoid-based pesticides due to their potential impact on honey bee populations.[60]

Other Factors

Other reported theories include the effects of shifting spring blooms and earlier nectar flow associated with broader global climate and temperature changes,[61] the effects of feed supplements that are produced from transgenic or genetically modified crops, such as high-fructose corn syrup,[62] and also the effects of cell phone transmissions and radiation from power lines that may be interfering with a bee's navigational capabilities.[63] The contributions of these possible factors have not been substantiated by evidence examined by the key researchers of this issue.[64]

Other Related Events

In late 2008, beekeepers in some states began to raise concerns about live bee imports from Australia. Specifically, these concerns centered on reports that the Asian honey bee, *Apis cerana*, had been introduced to Australia. Asian honey bees, found in southeast Asia, are considered an invasive species of bees and are known to carry a mite (*tropilaelaps clarae*). Neither the Asian honey bee nor the mite are known to be present in the Western Hemisphere. However, U.S. beekeepers have expressed concerns that bee imports could result in the introduction of mites and other diseases and further contribute to stressors already facing domestic species.[65] In November 2008, Australia notified APHIS of an incursion of Asian honey bees in the Cairns, Queensland, area of northeastern Australia, and Australia voluntarily stopped issuing export certificates to ship honey bees to the United States. In response, APHIS required that all honeybee exports be derived from colonies 100 miles away from any find of Asian honey bees known to have occurred in the last two years. As of late December, Australian shipments of honey bees to the United States have resumed. APHIS's decision to resume trade was based on data provided by Australia indicating that the areas outside the quarantine zone in Cairns are free the Asian honey bee and of the mites of concern.[66] Live bees (only queens with attendants and package bees) are allowed for import into the United States from Canada, Australia, and New Zealand. Imports from other countries are restricted. Certain import requirements apply along with general restrictions regarding the transit of live honey bees, bee byproducts, and bee equipment.[67] Items that are not allowed for transit include imports of whole colonies in hives, used beekeeping equipment, and pollen. Restrictions apply on beeswax for beekeeping and honey for bee feed, and require special treatment.[68]

ISSUES FOR CONGRESS

Committee Actions and Hearings

During the 110th Congress there were three House subcommittee hearings on honey bee colony declines and concerns about CCD. The House Subcommittee on Horticulture and

Organic Agriculture held two hearings: one in March 2007 and a second in June 2008. The Subcommittee on Fisheries, Wildlife, and Oceans of the House Committee on Natural Resources held a hearing in June 2007 on the role of pollinators in ecosystem health, which also addressed concerns about bee colony declines. In the Senate, in April 2008, Chairwoman Barbara Boxer and other members of the Senate Environment and Public Works Committee hosted a briefing on pollinators and their role in agricultural security.

Policy options discussed at these congressional hearings and briefings focused on the need for increased federal funding for multi-disciplinary research and monitoring to document changes in pollination reserves, as well as additional technical support and assistance for beekeepers. Additional research funding would help support USDA's research efforts and those at its laboratories located in Arizona, Louisiana, Maryland, Texas, and Utah.[69] Other recommended options included expanding crop insurance to include beekeepers and honey producers; providing a one-time payment for incurred losses; improving existing USDA conservation programs to better prevent habitat loss and sustain wildlife populations; emphasizing the importance of pollinator diversity and sustaining wild and native pollinator species; developing or improving existing federal and state best management practices for beekeepers; improving regulatory enforcement to prevent misuse of agricultural chemicals; and continuing the current marketing loan program for honey.

2008 Farm Bill

In May 2008, Congress enacted the 2008 farm bill (P.L. 110-246), which, among other things, provided additional funding for research and conservation programs addressing honey bees and pollinators. The law reflects provisions that were included in both the House- and Senate-passed versions of the farm bill, which addressed honey bees and pollinators as part of their conservation, specialty crop, research, and miscellaneous title provisions.

Conservation Provisions

The conservation title of the 2008 farm bill included language that broadly encourages habitat development and protection among the administrative requirements for native and managed pollinators under USDA's conservation programs (Section 2708), and ensures that USDA's conservation technical assistance includes standards that account for native and managed pollinators (Section 2706). These provisions could broaden the focus of USDA's farm conservation programs to include pollinator habitats and habitat improvement among their goals, as well as require USDA to review its conservation practice standards with respect to managed and native pollinators.

Research Provisions

The research title of the 2008 farm bill identified pollinator protection among its so-called high-priority research and extension areas (Section 7204). It provided for research and extension grants (1) to survey and collect data on bee colony production and health; (2) to investigate pollinator biology, immunology, ecology, genomics, and bioinformatics; (3) to conduct research on various factors that may be contributing to or associated with colony collapse disorder and other serious threats to the health of honey bees and other pollinators, including parasites and pathogens of pollinators, and the sublethal effects of insecticides, herbicides, and fungicides on honey bees and native and managed pollinators; (4) to develop mitigative and preventative measures to improve native and managed pollinator health; and (5) to promote the health of honey bees and native pollinators through habitat conservation and best management practices. For this provision, the 2008 farm bill authorized appropriations for grants at $10 million annually for FY2008-FY2012.

The research provisions also directed USDA to increase its capacity and infrastructure to address colony collapse disorder and other long-term threats to pollinator health (including hiring additional personnel) and to conduct research on colony collapse disorder and other pollinator issues at USDA's facilities. Annual appropriations were authorized at $7.25 million (FY2008- FY2012), with another $2.75 million annually (FY2008-FY2012) for honey bee pest and pathogen surveillance. The 2008 farm bill also directed USDA to submit an annual report to Congress on its response to CCD, indicating that the report should investigate the cause(s) of honey bee colony collapse and recommend appropriate strategies to reduce colony loss.

Insurance and Disaster Provisions

Other provisions in the 2008 farm bill supported pollinators through the bill's crop insurance and other disaster assistance provisions. One such provision identifies honey farms as possible beneficiaries of the bill's supplemental agricultural disaster assistance (Section 12033); another provision provides contracts for additional policies and studies to carry out research and development regarding insurance policies that cover loss of bees (Section 12023). Since enactment of the farm bill, USDA has created the Emergency Assistance for Livestock, Honey Bees, and Farm-Raised Fish Program (ELAP).[70] This program, administered by USDA's Farm Service Agency (FSA), provides disaster assistance for honey bee producers. The ELAP provides emergency relief to producers of livestock, honey bees, and farm-raised fish to aid in the reduction of losses because of disease, adverse weather, or other conditions, such as blizzards and wildfires. Eligible honey bee producers—those who incur physical losses of honey bees and honey bee hives because of colony collapse disorder—must provide documentation, and/or a certification that the loss of honey bees was due to CCD, from one or more of the following: registered entomologist; cooperative extension specialist; and/or land grant university. Additional information on this program is available at USDA's website[71] and at state county FSA offices.[72]

Other Provisions

The 2008 farm bill also contained provisions that generally support honey production. These include, for example, provisions pertaining to the National Honey Board (Section 10401-10402); provisions covering rates for marketing assistance loans for certain commodities, including honey (Section 1202); and provisions covering certain nutrition title provisions (such as Section 4231).

USDA Actions and Funding

USDA's Action Plan

USDA released its initial action plan for addressing CCD in July 2007. USDA's action plan focuses on improving coordination and redirecting existing resources and research for mitigation and prevention, including education and outreach, as well as expanding research and diagnostic resources to prevent future losses, working with the land grant universities. It also coordinates activities across three USDA agencies: the Agricultural Research Service (ARS), the Animal and Plant Health Inspection Service (APHIS), and the National Institute of Food and Agriculture (NIFA). USDA's focus on expanded research is consistent with the approach taken in the most recently introduced congressional bills and with recommendations by the American Honey Producers Association and the American Beekeeping Federation.[73]

Under the plan, USDA would (1) conduct surveys and collect data on bee health; (2) analyze bee samples for pests, disease-causing pathogens, pesticide exposure, and other factors; (3) conduct controlled experiments to identify factors affecting bee health, including potential causes of colony collapses; and (4) develop best management practices and guidelines to improve general bee health and reduce susceptibility to colony collapses and other disorders among both honey bees and non-*Apis* bees.[74] Aspects of USDA's action plan were presented at a hearing before the Subcommittee on Fisheries, Wildlife, and Oceans of the House Committee on Natural Resources in June 2007.[75]

USDA's 2007-2008 Progress Report

In June 2009, USDA published its 2007-2008 progress report on ongoing and intended future research efforts related to honeybees and CCD, following on the action items in its action plan. As outlined in USDA's progress report, prior study of the numerous possible causes for CCD has led researchers to further examine the hypothesis that CCD may be "a syndrome caused by many different factors, working in combination or synergistically."[76] Accordingly, future study will "focus increasingly on combinations and synergistic effects of factors in causing CCD."[77] The progress report provides detailed information on the status of ongoing research under each of the four elements of USDA's action plan, including survey and (sample) data collection, analysis of existing samples, research to identify factors affecting honey bee health, and mitigative and preventive measures. The progress report is available at USDA's CCD website at http://www.ars.usda.gov/is/br/ccd /ccd_ progress report.pdf.

Available USDA Research Funding

Funding for honey bee and CCD research at USDA's ARS has increased, following enactment of the 2008 farm bill (P.L. 110-246) and also FY2009 and FY2010 appropriations (P.L. 111-8 and P.L. 111-80, respectively), which, among other things, provide additional funding for research and conservation programs addressing honey bees and pollinators. Total ARS funding for honey bee and CCD research has been as follows:[78]

- FY2007—$7,675,000
- FY2008—$7,798,000
- FY2009—$8,290,000
- FY2010—$9,790,000 (includes $1.5 million increase for CCD research)

ARS also has an "Area-wide Project on Bee Health," which consists of temporary funding of $670,000 in FY2008 and will continue for at least four additional years at approximately $1 million per year.[79] Additional funding is available to USDA's NIFA, and includes combined research on honey bees, funding specific to CCD and bee health, and funding for various research labs and grants. Recently, emerging issues grants were awarded to Penn State University and the University of Georgia to study the effects of pesticides, pathogens, and miticides on pollinator populations.[80]

End Notes

[1] D. vanEngelsdorp1, J. Hayes, and J. Pettis, "Preliminary Results: A Survey of Honey Bee Colonies Losses in the U.S. Between September 2008 and April 2009," May 19, 2009, http://www.beealert.info/.

[2] Other known animal pollinators are stingless bees, bumble bees, other bees, wasps, hover flies, other flies, beetles, thrips, ants, butterflies, moths, bats, hummingbirds, and other birds.

[3] M. R. Berenbaum, University of Illinois, Statement before the Subcommittee on Horticulture and Organic Agriculture, U.S. House of Representatives, March 29, 2007, http://agriculture.house.gov/testimony /110/h70329/ Berenbaum.pdf; J. Pettis, USDA's ARS, interview with University of Pennsylvania staff, January 23, 2007. Staple crops (wheat, corn, and rice) do not rely on insect pollination and are mostly wind pollinated.

[4] R. A. Morse and N. W. Calderone, *The Value of Honey Bees as Pollinators of U.S. Crops in 2000*, March 2000, Cornell University, http://www.masterbeekeeper.org/pdf/pollination.pdf. Other studies show a range of estimated values from $5.7 billion to $19.0 billion (see National Research Council, *Status of Pollinators in North America*, 2006).

[5] Pollinator Partnership, "Helping the earth by supporting pollinators," 2009 publication.

[6] Ibid. Another study found that pollinators are essential for the production of some U.S.-grown crops, particularly macadamia nuts, squash, and pumpkins. A.M. Klein, et. al, "Importance of pollinators in changing landscapes for world crops," *Proceedings of the Royal Society B: Biological Sciences*, Vol. 274, No. 1608, February 7, 2007.

[7] Some "spillover" pollination occurs, including pollination from colonies owned by part-time beekeepers and hobbyists, or pollination of adjacent fields from commercial hives.

[8] Based on honey production statistics. USDA, *2007 Census of Agriculture*, Table 31, http://www.agcensus. usda.gov/ Publications/2007/Full_Report/Volume_1,_Chapter_1_US/st99_1_029_031.pdf.

[9] USDA, *Honey*, February 2009, http://usda.mannlib.cornell.edu/usda/current/Hone/Hone-02-27-2009.pdf.

[10] M. Burgett, *1999 Pacific Northwest Honey Bee Pollination Survey*, Oregon State University.

[11] USDA, CCD Steering Committee, "Colony Collapse Disorder Action Plan," June 20, 2007, at http://www.ars.usda.gov/is/br/ccd/ccd_actionplan.pdf; and USDA, "Questions and Answers: Colony Collapse Disorder," http://www.ars.usda.gov/News/docs.htm? docid=15572.

[12] See, for example, M. A. Aizen and L. D. Harder, "The Global Stock of Domesticated Honey Bees is Growing Slower than Agricultural Demand for Pollination," *Current Biology*, May 2009.

[13] National Academy of Sciences, National Research Council, *Status of Pollinators in North America*, 2006.

[14] D. vanEngelsdorp et al., "Fall Dwindle Disease: Investigations into the Causes of Sudden and Alarming Colony Losses Experienced by Beekeepers in the Fall of 2006," December 15, 2006.

[15] National Academy of Sciences, National Research Council, *Status of Pollinators in North America*, 2006.

[16] National Academy of Sciences, National Research Council, *Status of Pollinators in North America*, 2006; Interview with Maryann Frazier, Senior Extension Agent, Pennsylvania State University, January 28, 2007.

[17] R. A. Morse and N. W. Calderone, *The Value of Honey Bees as Pollinators of U.S. Crops in 2000*, March 2000, Cornell University, http://www.masterbeekeeper.org/pdf/pollination.pdf.

[18] Similar conditions have been termed autumn collapse, May disease, spring dwindle, disappearing disease, and fall dwindle disease.

[19] Pollinator Partnership, "Helping the earth by supporting pollinators," 2009 publication. Estimates attributed to Jeffrey S. Pettis at USDA's Agricultural Research Service (ARS).

[20] Interview with Maryann Frazier, Senior Extension Agent, Pennsylvania State University, January 28, 2007; vanEngelsdorp et al., "Fall Dwindle Disease: Investigations into the Causes of Sudden and Alarming Colony Losses Experienced by Beekeepers in the Fall of 2006," December 15, 2006.

[21] MAAREC, "Colony Collapse Disorder," http://maarec.cas.psu.edu/FAQ/FAQCCD.pdf.

[22] D. vanEngelsdorp, R. Underwood, D. Caron, and J. Hayes Jr., "An Estimate of Managed Colony Losses in the Winter of 2006-2007: A Report Commissioned by the Apiary Inspectors of America," *American Bee Journal*, July 2007, http://www.ento.psu.edu /MAAREC/CCDPpt/CCDJuly07ABJArticle-1.pdf. Based on a survey of

beekeepers that included 384 respondents representing 153,000 managed bee colonies located in AR, FL, GA, MD, MI, MS, MT, NM, ND, OH, PA, SD, TN, and WI.

[23] Estimated at the 95% confidence interval.

[24] These statistics may have been misrepresented in the popular press, which often state that 25% of the nation's 2.4 million colonies have been lost (citing the AIA survey as its source).

[25] Other reports indicate that the 2007 AIA survey found 30% colony losses.

[26] C. Henderson, J. Bromenshenk, L. Tarver, and D. Plummer, "National Honey Bee Loss Survey," June 2007, http://www.beealert.info/. Survey based on 625 responses received from operations in the U.S. and Canada, covering 43 states and five provinces. The majority (about two-thirds) of beekeepers surveyed were at smaller operations with less than 100 colonies, with another one-third of surveyed operations roughly split among operations with 100-1,000 colonies, 1,000-10,000 colonies, and more than 10,000 colonies.

[27] E. Burdick and D. M. Caron, MAAREC Beekeeper Survey, University of Delaware, http://maarec. cas.psu.edu/pdfs/ MAARECSurveyPub.pdf.

[28] D. vanEngelsdorp, J. Hayes, and J. Pettis, "Preliminary Results: A Survey of Honey Bee Colonies Losses in the U.S. Between September 2008 and April 2009," May 2009, http://www.beealert.info/. Survey based on about 20% of all U.S. colonies.

[29] Ibid.

[30] Jeff Pettis, "Colony Collapse Disorder Affecting Honey Bee (*Apis mellifera*) colonies," October 2008 presentation, http://www.epa.gov/pesticides/ppdc/2008/oct2008/session7-ccd.pdf. Survey based on 22 operations, managing 10% of U.S. colonies, and AIA surveys.

[31] D. vanEngelsdorp, D. Cox Foster, M. Frazier, N. Ostiguy, and J. Hayes, "Fall Dwindle Disease: Investigations into the Causes of Sudden and Alarming Colony Losses Experienced by Beekeepers in the Fall of 2006," December 2006.

[32] D. vanEngelsdorp et al., "Fall Dwindle Disease: Investigations into the Causes of Sudden and Alarming Colony Losses Experienced by Beekeepers in the Fall of 2006," December 15, 2006; published interview with Maryann Frazier, Penn State University, January 28, 2007, at http://podcasts.psu.edu/taxonomy/term/62; and Jeff Pettis, "Colony Collapse Disorder Affecting Honey Bee (*Apis mellifera*) colonies," October 2008 presentation.

[33] USDA, "Questions and Answers: Colony Collapse Disorder," http://www.ars.usda.gov /News/docs.htm?docid= 15572.

[34] Most queens are purchased from suppliers in Florida, California, Texas, Georgia, and Hawaii, or from suppliers in Canada, Australia, and New Zealand.

[35] D. Cox Foster, Pennsylvania State University, Statement before the Subcommittee on Horticulture and Organic Agriculture, U.S. House of Representative, March 29, 2007, http://agriculture.house.gov/testimony /110/h70329/ CoxFoster.pdf.

[36] D. vanEngelsdorp et al., "Fall Dwindle Disease: Investigations into the Causes of Sudden and Alarming Colony Losses Experienced by Beekeepers in the Fall of 2006," December 15, 2006; and C. Rexrod, USDA's ARS, Statement before the Subcommittee on Horticulture and Organic Agriculture, U.S. House of Representative, March 29, 2007, http://agriculture.house.gov/testimony/110/h70329/Rexroad.pdf.

[37] USDA, *Colony Collapse Disorder Progress Report*, CCD Steering Committee, June 2009, http://www.ars.usda.gov/ is/br/ccd/ccdjrogressreport.pdf. Also see D. vanEngelsdorp et al., "Colony Collapse Disorder: A Descriptive Study," *Proceedings of the National Academy of Sciences*, Vol. 4, Issue 8, August 2009.

[38] Ibid.

[39] USDA, "Questions and Answers: Colony Collapse Disorder," http://www.ars.usda.gov/News /docs.htm?docid= 15572.

[40] Published interview with Maryann Frazier, Penn State University, January 28, 2007, at http://podcasts.psu.edu/ taxonomy/term/62; and MAAREC, "Colony Collapse Disorder," http://maarec.cas.psu.edu/FAQ/FAQ CCD.pdf.

[41] USDA, CCD Steering Committee, "Colony Collapse Disorder Action Plan," June 20, 2007, at http://www.ars. usda.gov/is/br/ccd/ccd_actionplan.pdf.

[42] D. L. Cox-Foster et al., "A Metagenomic Survey of Microbes in Honey Bee Colony Collapse Disorder," *Science*, September 6, 2007, http://www.sciencemag.org/cgi/content /abstract/318/5848/283.

[43] USDA, "Questions and Answers: Colony Collapse Disorder," http://www.ars.usda.gov/News /docs.htm?docid= 15572. The study also found IAPV in honey bees from Australia that had been imported into the United States,

as well as in royal jelly imported from China. Further studies challenge the idea that IAPV is a recent introduction from imported bees.

[44] USDA press release, "Genetic Survey Finds Association Between CCD and Virus," September 2007, http://www.ars.usda.gov/is/pr/2007/070906.htm; and MAAREC statement, "Colony Collapse Disorder and Israeli Acute Paralysis Virus," http://maarec.cas.psu.edu/CCDPpt /CCDbeekeepersStatementIAPV.pdf.

[45] K. Ramanujan, "Parasites, pathogens and pesticides called possible suspects in honeybee decimation," *Cornell Chronicle*, Cornell University, May 17, 2007, http://www.news. cornell.edu/Chronicle/07/05_17_07.pdf.

[46] C. Henderson, J. Bromenshenk, L. Tarver, and D. Plummer, "National Honey Bee Loss Survey," June 2007, http://www.beealert.info/.

[47] D. Cox Foster, Pennsylvania State University, and C. Rexrod, USDA's ARS, statements before the Subcommittee on Horticulture and Organic Agriculture, U.S. House of Representatives, March 29, 2007, http://agriculture.house.gov/testimony/110/h70329/CoxFoster.pdf and http://agriculture.house.gov /testimony /110/h70329/Rexroad.pdf; and published interview with Jerry Hayes, Florida's Department of Agriculture, Apiary Section, March 2, 2007.

[48] M. Higes et al., "Honeybee colony collapse due to *Nosema ceranae* in professional apiaries," *Environmental Microbiology Reports*, Vol. 1, Issue 2, February 2009.

[49] Frazier, M., et. al, "What Have Pesticides Got To Do With It?" *American Bee Journal*, June 2008, http://maarec.cas.psu.edu/CCDPpt/WhatPesticidesToDoWithItJune08ABJ.pdf.

[50] S. Williams, "The Case of the Missing Bees," *Penn State Agriculture Magazine*, Winter/Spring 2008, http://aginfo.psu.edu/psa/08WinSpr/bees.html.

[51] National Academy of Sciences, NRC, *Status of Pollinators in North America*, 2006.

[52] D. Cox-Foster and D. vanEngelsdorp, "Solving the Mystery of the Vanishing Bees," *Scientific American*, March 31, 2009; Northwest Coalition for Alternatives to Pesticides, "Imidacloprid, Fact Sheet," *Journal of Pesticide Reform*, Spring 2001, http://www.pesticide.org/imidacloprid.pdf; and Apiculteurs de France, "Composite Document of Present Position Relating to Gaucho, Sunflower and Bees," http://www.beekeeping.com /articles/us/gaucho manifestationjaris_us.htm.

[53] Joe Cummins, "Neoniccotinoid insecticides used in seed dressing may be responsible for the collapse of honeybee colonies," April 24, 2007, http://www.organicconsumers.org /articles/article_4972.cfm.

[54] For example, see EPA's fact sheet on clothianiden, issued May 3002, http://www.epa.gov/opprd001/factsheets/ clothianidin.pdf

[55] D. vanEngelsdorp et al., "Fall Dwindle Disease: Investigations into the Causes of Sudden and Alarming Colony Losses Experienced by Beekeepers in the Fall of 2006," December 15, 2006; published interview with Jerry Hayes, Florida's Department of Agriculture, Apiary Section, March 2, 2007.

[56] Organic Consumers Association, "Honey Bee Health & Colony Collapse Disorder," http://www. organicconsumers.org/bees.cfm.

[57] NRDC press release, "Big Win for Bees: Judge Pulls Pesticide" December 29, 2009, http://www.nrdc.org/media/ 2009/091229.asp.

[58] NRDC press release, "EPA Buzz Kill: Is the Agency Hiding Colony Collapse Disorder Information?" August 18, 2008, http://www.nrdc.org/media/2008/080818a.asp.

[59] "German Coalition Sues Bayer Over Pesticide Honey Bee Deaths," *Environmental News Service*, August 25, 2008, http://www.ens-newswire.com/ens/aug2008/2008-08-25-01.asp.

[60] BNA's *International Environmental Law Committee Newsletter*, vol. 11, no. 1, February 2009, http://www. abanet.org/environ/committees/intenviron/newsletter/feb09/IELC_Feb09.pdf.

[61] W. Esaias, "Honey Bees, Satellites and Climate Change," presentation at the Library of Congress, April 3, 2007.

[62] See, for example, Louise A. Malone and Minh-Hà Pham-Delègue, "Effects of transgene products on honey bees (Apis mellifera) and bumblebees (Bombus sp.)," *Apidologie*, 32, 2001, http://www. hortresearch.co.nz/files/science/ gmimpacts/m1403malone.pdf; also research conducted by Hans-Hinrich Kaatz, University of Halle, Germany, cited at http://www.sierraclub.org/biotech/references.asp.

[63] Reportedly, this theory originated with initial research conducted in 2003 by J. Khun and H. Stever of Landau University in Germany.

[64] Statements and expert testimony at a public hearing of the U.S. House of Representatives, Subcommittee on Horticulture and Organic Agriculture, March 29, 2007.

[65] CRS communications with USDA APHIS on December 23, 2008. Also see, e.g., Kim Flottum, "Why the U.S. Should Stop Importing Bees from Australia," December 15, 2008, at http://www.thedailygreen.com /environmentalnews/blogs/bees/australian-bee-imports-88121501?src=rss.

[66] CRS communications with USDA APHIS on December 23, 2008.

[67] For information, see USDA APHIS import requirements: "Regulated Organism and Soil Permits, Honey Bees and Other Bees," at http://www.aphis.usda.gov/plant_health /permits/organism/bees/index.shtml.

[68] See 7 CRF Part 322, Subpart E, "Importation and Transit of Restricted Articles."

[69] There were reports in 2007 that the University of California at Davis was considering revitalizing its honey bee research program by hiring a bee breeder and geneticist and renovating the biology facility ("News Briefs," *AgriPulse*, Vol. 3, No. 20, May 16, 2007).

[70] USDA's final rule became effective in September 2009: 74 *Federal Register* 175: 46665-46683, http://edocket.access.gpo.gov/2009/pdf/E9-21906.pdf. See CRS Report RS21212, *Agricultural Disaster Assistance*.

[71] See, for example, USDA, "Documenting Losses Under the Emergency Assistance for Livestock, Honey Bees, and Farm-Raised Fish Program (ELAP)," http://www.fsa.usda.gov /Internet/FSA_Notice/ldap_2.pdf.

[72] For information on individual county offices, see http://offices.sc.egov.usda.gov/locator/app.

[73] R. Addee, American Honey Producers Association, Statement before the Subcommittee on Horticulture and Organic Agriculture, House Committee on Agriculture, March 29, 2007, http://agriculture. house.gov/testimony/110/h70329/ Adee.pdf; D. Weaver, American Beekeeping Federation, Inc., Statement before the Subcommittee on Fisheries, Wildlife and Oceans, House Committee on Natural Resources, June 26, 2007, http://resourcescommittee. house.gov/images/Documents/20070626 /testimony_weaver.pdf.

[74] USDA, CCD Steering Committee, "Colony Collapse Disorder Action Plan," June 20, 2007, http://www.ars.usda.gov/ is/br/ccd/ccd_actionplan.pdf.

[75] K. Hackett, USDA, Statement before the Subcommittee on Fisheries, Wildlife and Oceans, House Committee on Natural Resources, June 26, 2007, http://resourcescommittee.house. gov/images/Documents/20070626/ testimony_hackett.pdf.

[76] USDA, Colony Collapse Disorder Progress Report, CCD Steering Committee, June 2009, http://www. ars.usda.gov/ is/br/ccd/ccd_progressreport.pdf. This is the first annual report mandated by the 2008 farm bill (Sec. 7204 (h) (4)) on Honey Bee Colony Collapse Disorder (CCD).

[77] Ibid.

[78] CRS communication with USDA personnel, December 18, 2009.

[79] CRS communication with USDA personnel, December 4, 2008; Jeff Pettis, "Colony Collapse Disorder Affecting Honey Bee (*Apis mellifera*) Colonies," October 2008 presentation, http://www.epa.gov/pesticides/ppdc /2008/oct2008/ session7-ccd.pdf.

[80] Statements by Jeff Pettis, USDA, to Senate Environment and Public Works committee staff, April 9, 2008.

In: Bee Health: Factors, Analyses, and Research Progress ISBN: 978-1-62618-044-4
Editor: Roger Dunham

Chapter 3

COLONY COLLAPSE DISORDER: AN INCOMPLETE PUZZLE*

J. Kim Kaplan

When it comes to solving the puzzling syndrome known as "colony collapse disorder" (CCD), which has been attacking honey bee colonies since 2006, the best that can be said is that there is good news and bad news. The good news is that the rate of honey bee losses seems to have leveled off rather than continuing to increase. The bad news is that the cause or causes of CCD remain unclear.

In the United States, the problem surfaced in October 2006, when an increasing number of beekeepers began reporting losses of 30 to 90 percent of the hives in their apiaries with no apparent cause. The defining characteristic of CCD is the disappearance of most, if not all, of the adult honey bees in a colony, leaving behind honey and brood but no dead bee bodies. This definition has recently been revised to include low levels of Varroa mite and other pathogens, such as Nosema, as probable contributing factors.

For the last 3 years, self-reported surveys of beekeepers have found that winter colony losses have averaged about 30 percent, with around one-third of those losses ascribed to CCD, according to Jeff Pettis, research leader of the Bee Research Laboratory in Beltsville, Maryland, who heads up the Agricultural Research Service's CCD research effort. This compares to colony losses that were averaging 15-20 percent before CCD.

"The faint good news in the survey numbers is that the CCD problem does not seem to be getting worse," Pettis says. "But—and this is a big 'but'— 33-percent losses each year are probably not economically sustainable for commercial beekeeping operations."

While many possible causes for CCD have been proposed, reported, and discussed—both in the scientific literature and popular media—no cause has been proven. (See sidebar, page 7.)

"We know more now than we did a few years ago, but CCD has really been a 1,000-piece jigsaw puzzle, and the best I can say is that a lot of pieces have been turned over. The

* This article appeared in the July 2012 issue of Agricultural Research, a publication of United States Department of Agriculture.

problem is that they have almost all been blue-sky pieces—frame but no center picture," Pettis explains.

In Beltsville, Maryland, ARS entomologist Jay Evans inspects a comb of honey bees for signs of mites and brood disease.

Entomologist Jay Evans (left) and postdoctoral research associate Ryan Schwarz use a microscope to look at spores of the honey bee fungal parasite Nosema ceranae, which can replicate in cells lining the honey bee gut.

The bee lab's scientists have been looking for the cause or causes of CCD within four broad categories: pathogens; parasites, such as *Varroa* mites or *Nosema;* environmental stressors, such as pesticides or lack of nectar diversity; and management stressors. The researchers have been analyzing samples from healthy and CCD-struck colonies and applying a variety of stressors from the four groups to colonies in hopes of provoking a colony response that duplicates CCD.

"While a number of potential causes have been championed by a variety of researchers and interest groups, none of the causes has stood up to detailed scrutiny. Every time someone has proclaimed a potential smoking gun, further investigation has not been able to make the leap from correlation to cause-and-effect for one reason or another. Other times, there hasn't even been a scientific correlation," Pettis says.

ARS's research work, along with that of university and other scientists, "pretty well supports the idea that CCD is causedby multiple factors—possibly working individually, but more likely in combination," Pettis adds. "But we still can't say whether it's the same set of factors in every CCD incident or—if it *is* the same factors—that they are occurring in the same sequence in every case."

ARS scientists are studying the transfer of food or other fluids among members of a bee community through mouth-to-mouth feeding as a possible facilitator of colony collapse disorder.

Pathogens continue to stand out as one of the important puzzle pieces, according to ARS entomologist Jay Evans, also with the Bee Research Laboratory. He was part of a team that used genetic analyses to look for correlations between bee health and bee pathogens and activity levels of honey bee genes.

Two picornalike viruses—acute bee paralysis virus and Kashmir bee virus— along with deformed wing virus, black queen cell virus, and two species of *Nosema,* were found to be more abundant in CCD hives. Infection by multiple picornalike viruses could result in honey bees having reduced abilities to synthesize certain proteins, the lack of which would leave the bees more vulnerable to additional stresses like pesticides, nutritionproblems, or other pathogens—which sounds like a possible root cause of CCD, Evans points out.

When the researchers looked at the bees' turning on of detoxification and immune genes, which would have reflected exposure to either pesticides or disease, respectively, there was no significant difference between CCD and non-CCD colonies.

A honey bee, with pollen attached to its hind leg, pollinating a watermelon flower.

The team did find considerable differences between CCD hives on the west coast and the east coast. "Finding Kashmir bee virus in a hive was the best predictor of CCD in the western United States, while deformed wing virus, an unrelated RNA virus, was a better predictor in the East," Evans says.

Evans and Bee Research Laboratory colleague Judy Chen were also part of an international team that closely followed 29 European honey bee colonies, carefully monitoring for pathogens, parasites, and bee proteins. This study found that four factors appeared to be the best predictors so far of winter honey bee loss: presence of the microsporidian *Nosema ceranae;* levels of the protein vitellogenin, which strongly reflects the bee's protein status and plays an immune system role; *Varroa* mite infestation; and the presence of deformed wing virus, a virus often associated with *Varroa* mites.

Finding *Varroa* mites to be a good predictor of winter declines in this study was not really surprising, according to Evans, though other studies in the United States have not found as high a correlation between CCD and *Varroa* mites. "Even if *Varroa* mites themselves do not directly cause CCD, we know they can transmit multiple viruses to honey bees," Evans says, "and higher total pathogen levels (rather than infection by any specific pathogen) have been our best correlation with CCD so far. But should the pathogens be considered primary casual agents— multiple pathogens specifically causing CCD—or are they indirectly involved because they weaken bees, making these bees more vulnerable to something else we don't know yet?"

PESTICIDES AS CAUSE?

Pesticides—individually and in general—have been repeatedly nominated as a cause of CCD, often without direct scientific data to support the idea. In a pesticide survey conducted by U.S. Department of Agriculture and university scientists that analyzed wax, pollen, and bee samples for the presence of 121 different pesticides or their metabolites, the most commonly found pesticides were fluvalinate and coumaphos. While about 60 percent of the 259 wax and 350 pollen samples did show the presence of at least one systemic pesticide, almost all were found at levels well below what is considered lethal to honey bees.

There was no overall pattern of exposure among the samples for a specific pesticide or class of pesticides. The study did not look specifically at the pesticides as they might be related to CCD, but if a specific class of pesticides were involved, a pattern of residues should have been discernable, explains Pettis, who co-led the study. Not all pesticide impact is about directly killing honey bees, however. Sublethal doses of the pesticide imidacloprid—one of the neonicotinoid group of pesticides— were found to make honey bees more susceptible to the gut parasite *Nosema,* according to a study by Pettis and University of Maryland researchers Dennis vanEngelsdorp, Josephine Johnson, and Galen Dively. The researchers fed three generations of honey bee colonies either 5 or 20 parts per billion (ppb) of imidacloprid, which is used to protect a wide variety of crops and ornamentals from many different insects. The dosages used in the study were intentionally well below the levels that have been documented to kill honey bees after short-term exposure and reflected levels that have been measured in the environment. After the third generation, newly emerged adult bees from these colonies were exposed to spores of *N. apis* and *N. ceranae,* gut parasites that have been a growing problem for U.S. beekeepers since the 1990s. There was up to a fourfold increase inthe levels of *Nosema* in honey bees from the imidacloprid-exposed colonies, regardless of whether 5 or 20 ppb were fed. "While these increased *Nosema* levels were found in individual bees, there was no measurable impact at the colony level," Pettis says.

"Imidacloprid was chosen for this study because of its widespread use and beekeepers' concerns about it. But it was only found in 3 percent of the pollen samples checked in the pesticide-survey study, usually at very low levels, and no connection with CCD has ever been made scientifically," he adds. To better account for such sublethal impacts, Pettis is working as part of an international group of scientists and regulators to help the U.S. Environmental Protection Agency (EPA) develop recommendations for pesticide-testing guidelines that factor sublethal effects into test protocols. ARS scientists have already developed a sublethal-impact cage assay that has been provided to EPA. Could imidacloprid and *Nosema* togetherbe the cause ofCCD, as some claim? "This study did not look for nor establish any connection between either imidacloprid or *Nosema* and CCD," Pettis explains. "But the effect of the combination of imidacloprid and *Nosema* demonstrates that there are many complex interactions between stress factors that need to be considered in looking for a cause of CCD and high honey bee mortality in general."—By *J. Kim Kaplan,* ARS. *This research is part ofCropProduction, anARSnationalprogram (#305) described at www.nps.ars.usda.gov. To reach scientists mentioned in this article, contact Kim Kaplan, USDA-ARS Information Staff, 5601 Sunnyside Ave., Beltsville, MD 20705-5128; (301) 504- 1637, kim.kaplan@ ars.usda.gov.**

Many Suspects, but No Culprit

Colony Collapse disporder (CCD) has been a subject of interest in scientific journals and the popular media since the syndrome first appeared in 2006. Despite numerous and differing claims, nothing has actually been shown to be the cause of the problem.

Pathogens

One or more pathogens remain scientists' most likely choice as the cause or partial cause. But various viruses and bacteria have had highercorrelations with CCD in different parts of the United States and in different countries. Before any pathogen can be legitimately accepted as the cause, science must demonstrate that when it is introduced into a healthy colony, CCD results.

Parasites

A parasite is the other perennial suspect, either by itself or in combination with one or more pathogens. Nosema and Varroa mites remain high on the probable-cause list.

New Pests or Diseases

Some believe that a previously undiscovered or unidentified pest or pathogen is involved in CCD. But claims that such an agent has been identified have not held up scientifically so far.

Pesticides

There are many classes of pesticides to which honey bees can become exposed.

Among those that have been stamped with a "CCD cause" label are the neonicotinoids, like imidacloprid and clothianidin. One issue with making that link is the lack of a matching pattern between neonicotinoid residues in colonies and CCD outbreaks. France, which banned imidacloprid in 1999, and Germany, which along with France banned clothianidin in 2008, still have CCD problems.

Transportation Stresses from Migratory Beekeeping

Pollination-service beekeepers stack colonies on tractor-trailers and transport them thousands of miles during the growing season. For honey bees, orientation to their hive is vital, and being regularly relocated must be stressful. Additionally, moving hives around the country may spread diseases and pathogens as honey bees intermingle in the fields.

It is possible that such stresses play into CCD, but there is no scientific evidence of it at this time.

Monoculture

Wild honey bees forage on a wide variety of nectar sources. Honey bees used for commercial pollination are mostly limited to one crop at a time, and it is possible that they may suffer nutritional deficiencies that stress their immune systems.

Genetically Modified Crops

Genetically modified (GM) crops, most commonly Bt corn, have been offered up as the cause of CCD. But there is no correlation between where GM crops are planted and the pattern of CCD incidents. Also, GM crops have been widely planted since the late 1990s, but CCD did not appear until 2006. In addition, CCD has been reported in countries that do not allow GM crops to be planted, such as Switzerland. German researchers have noted in one study a possible correlation between exposure to Bt pollen and compromised immunity to Nosema.

High-Fructose Corn Syrup

Some researchers have attributed CCD to the practice of feeding high-fructose corn syrup (HFCS) to supplement bee colonies. But there are many reports of CCD occurring in the apiaries of beekeepers who do not feed HFCS. Others have suggested a possible connection with HFCS produced from genetically modified corn, combining two popular villains. But the simple management change of not feeding any HFCS does not stop CCD.

Global Climate Change

Weather changes, such as unusually warm winters, earlier springs, drought, and flooding, can lead to changes in flowering times. Plants may blossom early, limiting nectar and pollen supplies. But bees used for pollination contracts are moved to fields to coincide with flowering of crops. Still, some believe global warming is to blame, if only in part, for CCD.

Ozone

The level of the air pollutant ozone has been steadily dropping since the early 1990s. Since CCD did not appear until 2006, the timing doesn't match for ozone to be related.

Cell phones and cell phone towers The idea of cell phones causing CCD began with the misinterpretation of a study in which a cordless home phone, not a cell phone, was shown to have some impact on honey bee navigation. The study author has repeatedly stated that the phone he tested is nothing like a cell phone and has nothing to do with CCD. But the idea remains popular. One of the most recent "proofs," (published in Current Science in 2010) claimed evidence suggesting "that colonycollapse does occur as a result of exposure to cell phone radiations" while also reporting that the impact of cell phones in both of the test hives resulted in more bees staying in the hive longer—the exact opposite of the definition of CCD.*

In: Bee Health: Factors, Analyses, and Research Progress ISBN: 978-1-62618-044-4
Editor: Roger Dunham

Chapter 4

2011-2012 National Honey Bee Pests and Diseases Survey Report*

Karen Rennich, Jeff Pettis, Dennis Vanengelsdorp, Rachel Bozarth, Heather Eversole, Karen Roccasecca, Margaret Smith, Jennie Stitzinger, Michael Andree, Rob Snyder, Nathan Rice, Jay Evans, Vic Levi, Dawn Lopez and Robyn Rose

Executive Summary

The 2011 USDA Animal Plant and Health Inspection (APHIS) sponsored National Survey of Honey Bee Pests and Diseases was conducted in collaboration with the University of Maryland (UMD) and the USDA Agricultural Research Service (ARS) and with the cooperation of 34 states covering the timeframe from the spring of 2011 through the summer of 2012. This survey is in its third year and is the largest to date. The National Survey has expanded from a 3 state Pilot Survey in 2009 to a Limited National survey of 13 states in 2010. This expansion has allowed us to augment and extend the baseline pest and pathogen data collected from the previous 2 surveys. The primary focus of this survey was to verify the absence of exotic threats such as the parasitic mite, *Tropilaelaps* spp., *Apis cerana*, and Slow Paralysis Virus. Under current international trade agreements, the U.S. cannot deny import permits from other nations unless the exporting nation has a disease, parasite, or pest of honey bees that is not found in the U.S. Establishing the absence of threats to honey bee populations not thought to be present in the U.S. was the primary objective of this effort.

With exotic species as the primary target, a secondary objective was set to make use of the sampling by determining existing levels of other honey bee diseases and parasites known to be present in the U.S. This was also performed in the previous two surveys. The survey results are used to gauge the overall health of colonies, to create a baseline disease level, and to facilitate interpretation of ongoing and future epidemiological studies. This baseline data,

* This is an edited, reformatted and augmented version of the USDA Animal and Plant Health Inspection Service publication, dated 2012.

including historic data from research institutions such as the ARS Bee Research Laboratory (BRL) and other ongoing field sampling and management surveys, have been incorporated into a single database as part of the Bee Informed Partnership (www.beeinformed.org), a 5 year grant funded by USDA NIFA (National Institute of Food and Agriculture). The 2011-2012 National Survey effort included collection of samples from 34 states: Alabama, Arkansas, California, Colorado, Delaware, Florida, Georgia, Hawaii, Idaho, Illinois, Indiana, Iowa, Louisiana, Maryland, Michigan, Minnesota, Montana, Nebraska, New York, New Mexico, New Hampshire, New Jersey, North Carolina, North Dakota, Ohio, Pennsylvania, South Carolina, South Dakota, Tennessee, Texas, Utah, Virginia, West Virginia and Wisconsin. A total of 875 samples representing 7,000 colonies are expected at the completion of this survey. To date, approximately 94% of the live bee samples (for virus analysis) have been collected and analyzed and about 93% of the alcohol samples (for Varroa mite, Nosema and Tropilaelaps determination) have been delivered and processed. Also at this time, 80% of the reports to beekeepers and apiary inspectors have been distributed. A second full National Survey for 2012 was initiated early this summer with 32 participating states. Samples for this fourth survey have already begun to arrive.

The survey samples were analyzed for 10 known honey bee viruses, pests and pathogens. Slow Bee Paralysis Virus (SBPV), the only virus included in this year's testing that is not currently found in the U.S., was examined in all samples and no detection was made. No diseases or parasites of bees not already known to exist in the country were discovered. Two common honey bee viruses, Deformed Wing Virus (DWV) and Black Queen Cell Virus (BQCV), were found in all 34 states. In this 2011 survey, 9 samples out of 673 tested positive for *Nosema apis*, or about 1.3% of the samples. This *Nosema* species was not detected in the previous survey years; however, this may be attributed to the development and implementation of a more sensitive primer used in the Polymerase Chain Reaction (PCR) detection this year. For the third year, we saw no evidence of *Tropilaelaps* mites. Since honey bee tracheal mites (*Acarapis woodi*) were not detected in samples in 2009 nor in 2010, samples were not analyzed for this mite this year. Honey bee tracheal mites are known to exist in the country and our failure to find them may be the result of our sampling procedure. Varroa mites continued to be observed in all states with the exception of the Hawaiian Islands of Maui, Kauai and Molokai.

New to the survey this year was a pilot pollen pesticide survey where a minimum of 3 grams of pollen was collected from brood frames that was tested for 174 known pesticides. The pollen was collected from the same composite 8 colonies undergoing the standard survey sampling and sent to the USDA Agricultural Marketing Service (AMS) in Gastonia, NC for analysis. This pilot pesticide survey focused on 11 states: Alabama, California, Colorado, Florida, Idaho, Indiana, New York, South Dakota, Tennessee, Texas and Wisconsin. We received 99 samples from these states and of the 174 possible pesticides, only 35 were detected. This pollen survey will be expanded and will include each participating state in the 2012 National Survey.

This survey was designed to be representative of the managed honey bees across the broad geography of the United States. The survey was open to any state wishing to participate and beekeeper participation was completely voluntary and the beekeeper did not have to be present or assist with the sampling. The results can be considered as representative of the pests and pathogens present in the U.S.

INTRODUCTION

This 34 state USDA survey of honey bee pests and pathogens began in 2011 and will be completed in late summer 2012. Variability from 1) the climates when samples can be collected from active brood areas in northern states and 2) the time frame the states received their respective agreements, has caused a delay in the completion of the survey. We have received 94% of all samples required and it is anticipated that remainder will arrive in August 2012. Funding was provided by the USDA Animal and Plant Health Inspection Service (APHIS) and the survey was conducted in collaboration with the USDA Agricultural Research Service (ARS), and the University of Maryland (UMD). A total of 392 colonies to date were sampled from 49 apiaries in California (11 from migratory beekeepers who were in that state for pollination contracts and 38 from beekeepers originating from there). 25 samples incorporating 5,600 colonies were sampled from the following states: Alabama, Arkansas, Colorado, Florida, Georgia, Hawaii, Illinois, Indiana, Iowa, Louisiana, Maryland, Michigan, Minnesota, Montana, Nebraska, New York, New Jersey, North Carolina, North Dakota, Pennsylvania, South Carolina, South Dakota, Tennessee, Texas, Utah, Virginia, West Virginia and Wisconsin. Additional samples were and are still being received from Delaware, Idaho, New Mexico, New Hampshire and Ohio.

Survey Description

Survey kits were distributed to the participating states' Department of Agriculture offices in June 2011. This late date did not allow for some states to complete sampling in 2011 and they had to resume in the spring of 2012. Based on this experience, the 2012 survey kits were shipped to the states by late April and early May 2012. Apiary inspectors and agents conducted an aggregate sampling from previously identified commercial, migratory, and sideliner beekeepers with at least 8 colonies per apiary. In most cases apiaries consisted of at least 10 colonies. A single aggregate sample was collected from 8 randomly selected colonies per apiary per operation (APHIS US Honey Bee Survey Sampling Protocol). In each state, apiaries were chosen on a case by case basis with an attempt to give as close to an equal representation of the entire state as possible. Ideally, a state was sectioned into 4 quadrants with apiaries randomly chosen within a quadrant. When possible, ten queen producers were sampled. Of the remaining sampled apiaries, 1/2 were from migratory operations (move out of the state and return prior to sampling) and 1/2 were from stationary operations (only move within the state or do not move at all). Additional apiaries occurring near ports or other areas that could be considered high risk were also considered for sampling (APHIS US Honey Bee Survey Project Plan). The pollen samples were collected (APHIS US Honey Bee Survey Pollen Sampling Protocol) concurrently and from the same colonies in the apiary being sampled for the disease and pest survey.

Four distinct collection methods were used to sample each apiary. The first sample was a collection of live adult bees composed of 1/4 cup of bees (~ 150 bees) that were shaken off brood frames from each of the 8 sampled colonies. The live bees were deposited in a live bee shipping box containing a water source and hard sugar candy (fondant). This box was shipped the same day to the USDA/ARS in Beltsville, MD where it was immediately frozen at -80C until molecular testing could be performed. The molecular tests were performed with quantitativePCR techniques, outlined by Dr. Jay Evans at the USDA/ARS Bee Research Laboratory, to look for molecular evidence of known viruses and other pests (2006 and Honey Bee PCR Diagnostics). We have incorporated a new, high-performance chemistry in our molecular analysis this year that provides greater sensitivity, faster analysis times, and is more cost effective. The molecular tests were designed to detect the presence of the following:

1. Acute Bee Paralysis Virus (ABPV)
2. Deformed Wing Virus (DWV)
3. Israeli Acute Paralysis Virus (IAPV)
4. Chronic Bee Paralysis Virus (CBPV)
5. Black Queen Cell Virus (BQCV)
6. Slow Bee Paralysis Virus (SBPV)
7. Nosema ceranae
8. Nosema apis

The second sample of bees, consisting of 1/4 cup of bees from each of the 8 sampled colonies, originated from the same brood frames as the live bee sample. These bees were put into a bottle of alcohol for preservation. This alcohol sample was shipped to USDA/BRL for visual and microscopic analysis by University of Maryland technicians to quantify the following:

1. Nosema spp. spores
2. Varroa Mite loads
3. Apis cerana

The third sample was taken from anything dislodged from 'bumping' sampled brood frames over a collection pan. This technique was developed by Dr. Jeff Pettis and Dr. Dennis vanEngelsdorp and funded by APHIS as a quick and cost effective way to detect for the *Tropilaelaps* mite. The sample, also preserved in alcohol, included any mites, beetles and other hive debris filtered from bumping the brood frames. This sample was shipped to USDA/ARS Beltsville, MD and analyzed for the presence of the *Tropilaelaps* mite.

Finally, the fourth sample, taken only by those states participating in the Pilot pesticide/pollen survey, collected a minimum of 3 grams of dry pollen (not bee bread) from the same colonies, preferably in the same brood area, from the other three samples described above. These samples were placed in a tube, labeled and sent to USDA/ARS Bee Research Laboratory where they were refrigerated until the entire assortment of samples could be analyzed at USDA/AMS.

All participating beekeepers, as well as State Apiarist/Inspectors, received a single report for each sample taken. This report, revised after last year, contains considerably more information than previous reports. It is usually sent within 6 months of collection and details the analysis results for Varroa mite load, *Nosema* load, all six viruses and the species identification of *Nosema* (*N. apis* or *N. ceranae*) and notes the presence or absence of *A. cerana* and *Tropilaelaps* spp. It also includes the national prevalence for viruses at the date of the report as well as specific beekeeper percentile rankings of Varroa mite load, *Nosema* spore load and viral loads for the 2 most common viruses, Deformed Wing Virus (DWV) and Black Queen Cell Virus (BQCV).

Using the U.S. Postal Service, live bee shipments were made to USDA/ARS and percent survivability was tracked for all live bee shipments. The results of this analysis, previously proven to be a robust and suitable alternative for shipping bees on dry ice by the Pilot and Limited Survey, continued to work well. In some states, a small number of live bee samples were degraded, such that no molecular data could be obtained for these samples. Additionally, some live bee samples were lost in shipment. The lost live bee samples occurred in one sample from Hawaii, one from Pennsylvania and two from Montana.

Results

Pest Survey

The results of all analyses can be found in the Appendix. From the 778 alcohol samples analyzed for *Nosema* via microscopy, 442 (56.7%) has detectable spore loads. An examination of 95% CI surrounding the prevalence of *Nosema* infected apiaries suggest that prevalence was highest in 2009 (Figure 1); however, caution should be used when interpreting this data as the sampling frame changed between years. Of the samples in 2011 that tested positive for *Nosema*, 86 samples (19.5%) exceeded the threshold thought to cause damage (> 1 million spores per bee). Samples testing positive for *Nosema* infection had a mean *Nosema* spore load of 564,000 spores per bee, (Figure 2). Figures 3 and 4 illustrate *Nosema* prevalence and *Nosema* spore load from all 3 years of the survey on a monthly basis. Any month having less than 3 data points was not included in the monthly calculations. Figure 3 shows the classic seasonal decline in *Nosema* detection in the late summer and early fall in conjunction with a decrease in detectable spore load in those same months (Figure 4). *Nosema* levels typically appeared highest in late fall and early spring months and spring 2011 samples showed higher loads than last year. Data on treatments were not taken this year so there is no indication which, if any, of these colonies were treated for *Nosema*. We hope to include a short management survey, in conjunction with the Bee Informed Partnership, this year to capture this relevant and critical data.

Varroa mite prevalence for 2011 is similar to that observed in 2010 (92.4%) and 2009 (87.1%) and indicates that almost 91.8% of all 781 alcohol samples had at least one Varroa mite (Figure 5). While prevalence did not change over the tree years of survey, the overall mite load in infested apiaries has increased steadily since 2009 (Figure 6). Again, caution should be used when interpreting this data as the 2009 sampling time frame was not the same as in subsequent years. While the economic threshold for Varroa mites is seasonally and regionally specific, generally an average load of over 3 mites per 100 bees is of concern. Worryingly in 2011, 62.9% of the samples that tested positive for Varroa (450 out of 716) exceeded the lower threshold for possible damage to a colony from Varroa.

Figures 7 and 8 illustrate the dynamic nature of Varroa mite populations over the course of the year. Varroa mite levels were highest in the late summer and fall months and Figure 7 demonstrates 100% of the samples received in May 2012 had Varroa mites. Again, like *Nosema*, it is unknown whether the sampled apiaries treated for Varroa mites, but future surveys will attempt to gather that information.

Figures 9, 10, and 11 illustrate the viral prevalence profiles for the survey years 2009, 2010 and 2011 respectively. Four viruses were consistently tested for all 3 years and include Israeli Acute Paralysis Virus, Deformed Wing Virus, Acute Bee Paralysis Virus and Slow Bee Paralysis Virus (SBPV). The survey in 2009 reported the highest incidence of IAPV but 2010 saw the highest incidence of DWV and ABPV. In this survey year, DWV and BQCV, the two most ubiquitous viruses among honey bees, were detected in every state sampled.

The monthly prevalence of four commonly found viruses (IAPV, DWV, ABPV and BQCV) is provided in Figures 12, 13, 14 and 15. IAPV (Figure 12) and ABPV (Figure 14) illustrate seasonality in these viruses. In contrast, DWV (Figure 13) and BQCV (Figure 15) exhibit a more constant presence with a slight peak for BQCV seen in the late spring/early summer months. No monthly prevalence graphs are provided for SBPV as this virus was not detected and the data set was very limited for CBPV. It is anticipated that 7 viruses will be tested for in 2012 and include ABPV, BQCV, CBPV, DWV, IAPV, SBPV, and Kashmir Bee Virus (KBV).

Finally, this study found no evidence of Tropilaelaps or Apis cerana. Visual analysis of samples collected in alcohol did not detect a presence of this exotic Apis species and A. mellifera sub-races.

The molecular techniques employed in this survey are based on analysis of the RNA extracted from each sample. Therefore, our molecular identification focuses on detection of actively reproducing Nosema (vegetative stage), not dormant (spore stage) Nosema. Subsequently, it is possible that the samples examined by microscopy had detectable levels of Nosema (spores) while the molecular analysis quantifies active infection. This accounts for the difference in the PCR and microscopic detection of Nosema in these samples (Figure 16).

It should also be noted that nine N. apis detections were made this year, representing 1.3% of the samples. This is the first year since survey inception that N. apis has been found. A newly designed, more sensitive primer was used this year to detect Nosema apis. This would explain the detections for N. apis this year and not in previous years. 49 of 669 samples (7.3%) tested positive for N. cerana, which still appears to have largely replaced N. apis in the European honey bee (Apis mellifera) after migrating from its original host, A. cerana.

Pesticide Survey

A pilot survey of pesticides found in pollen sampled from hives was initiated this year to test the pollen sampling and shipping protocol. The pesticide survey included 11 states listed in the summary and they collected the required pollen in addition to their concurrent colony sampling for live bees and alcohol samples. The most prevalent pesticides include Coumaphos and its metabolites (detected in 39.4% of the samples), Fluvalinate (detected in 38.4% of the samples), Thymol and 2, 4 Dimethylphenyl (a metabolite of Amitraz), both detected in 27.3% of the samples and Chlorpyrifos (detected in 20.2% of the samples). The full set of results, grouped by their classification as an insecticide, herbicide or fungicide, is given in Figure 17. The level of detection (LOD), or the minimum amount that can be reliably detected, the national prevalence (%) seen by this limited survey, the average level detected (parts per billion or ppb) and the range of detection (ppb) are provided for those samples that tested positive for that specific pesticide. If a pesticide was detected once, a single value is given for the range and it is marked with an asterisk. With the additional information of pollen collected concurrently with the live bee and Nosema and Varroa mite samples, it may be possible to correlate colony health to in hive pesticide residue and there are plans to do this with the 2012 survey. The 2012 state survey will include 10 samples from each of the 33 states.

Conclusion

The increased sample size this year allows for the expansion of our database of pests and pathogens and places the collected data into a temporal context. The 2011 National Survey was expanded to capture some states not previously sampled and the sampling season was lengthened with the inclusion of more southern states. This allowed us to greatly increase viral and pest data for the winter months as shown by the prevalence graphs for the more common viruses. Varroa mite loads were seen to increase over the 3 years of the survey and it is unknown what this cause may be but possible explanations include fewer beekeepers treating, mites becoming resistant to available treatments or an improved snapshot of the actual mite infestation across the country. Results that will be monitored this year include Varroa mite loads to determine if the increasing trend continues and what, if any treatments, are being applied to the sampled colonies. Additional data will be collected on BQCV as not much is known about this very common and established virus. Observing the dynamics and possible seasonality between *N. apis* and *N. ceranae*, in conjunction with treatment data we hope to collect will also be core to the 2012 survey. By gathering yearly, sequential samples from a growing number of states, we may be able to see trends and patterns that relate to colony health. The survey does provide strong evidence that *Tropilaelaps*, Slow Paralysis Virus and *Apis cerana* are not present in the U.S.

Appendix

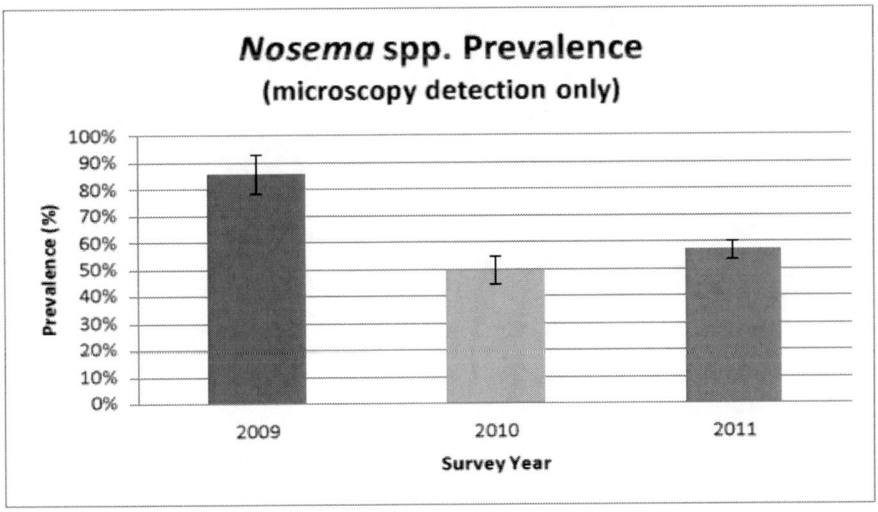

Figure 1. Nosema spp. prevalence over 3 years of survey. (95% Confidence Intervals shown).

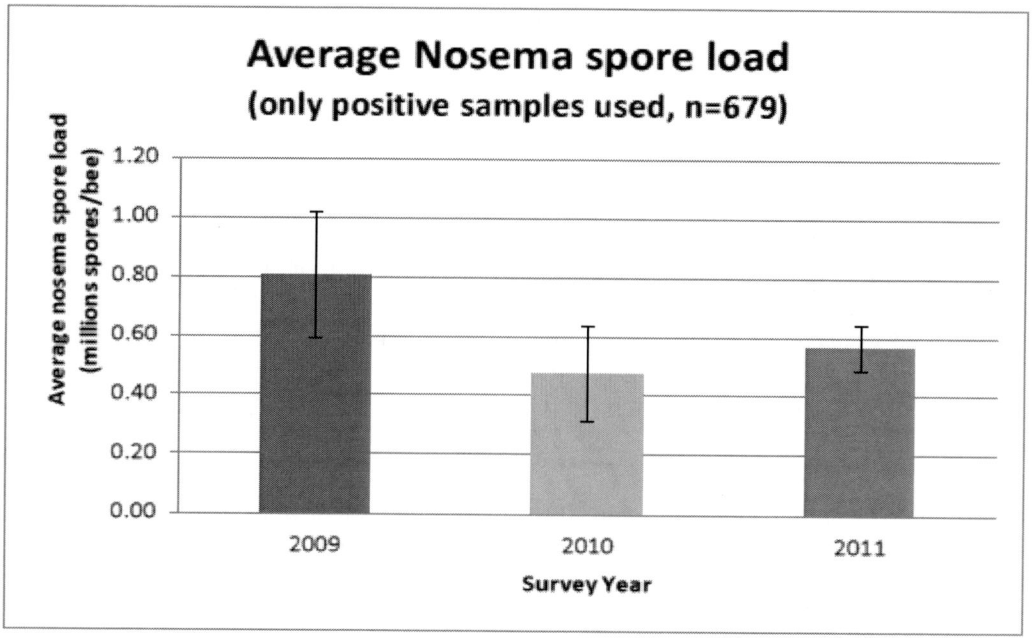

Figure 2. Nosema spp. spore load over 3 years of survey. (95% Confidence Intervals shown).

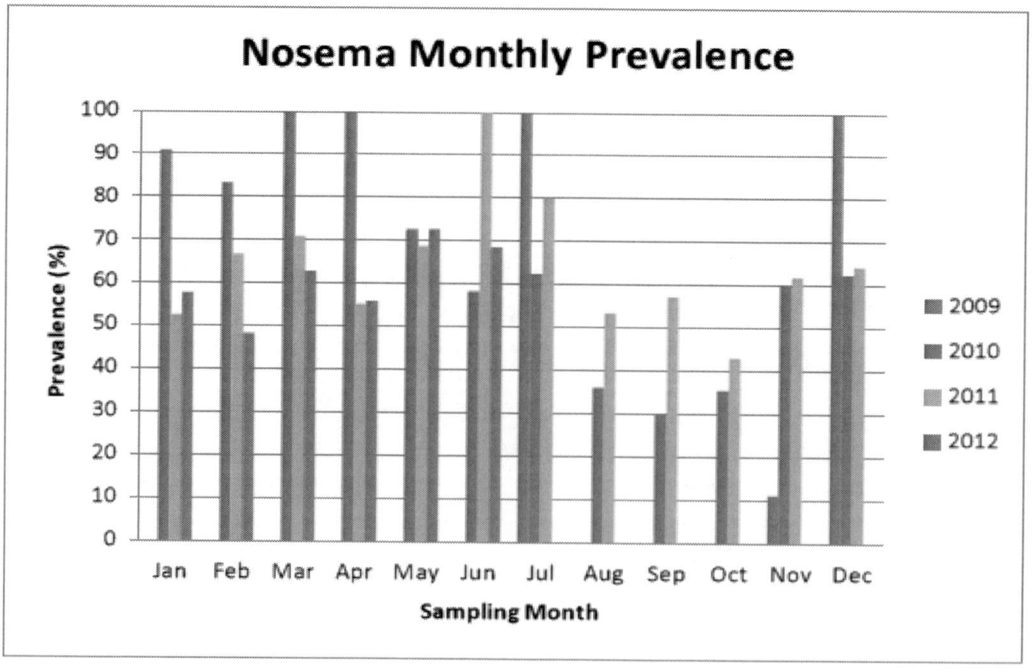

Figure 3. Monthly prevalence for Nosma spp. (95% Confidence Intervals shown).

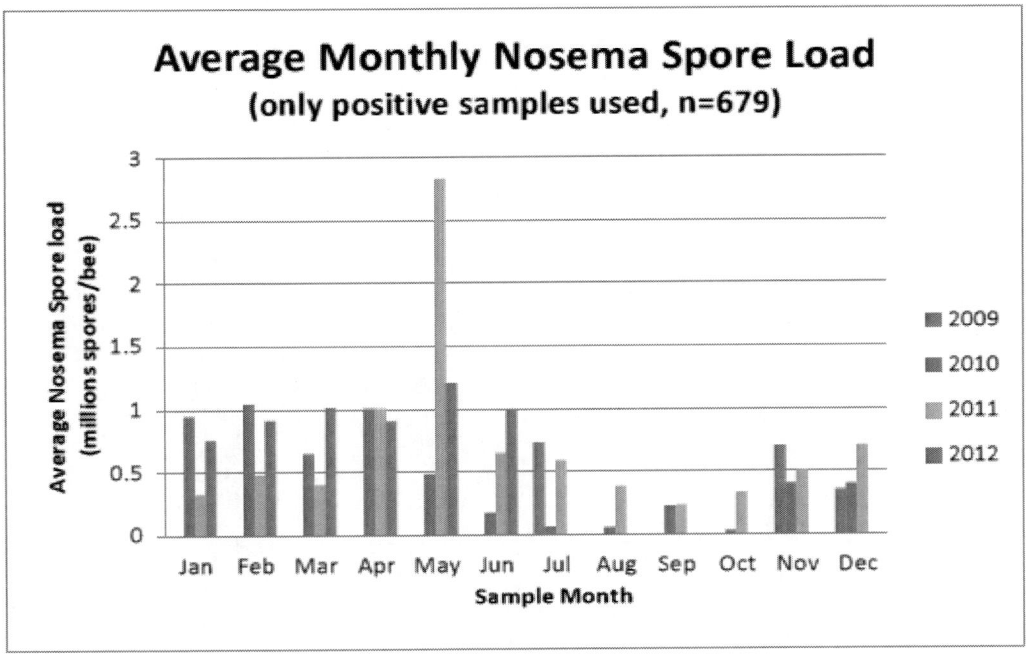

Figure 4. Average Monthly Nosema spore load. (for samples testing positive by microscopic spore count) (95% Confidence Intervals shown).

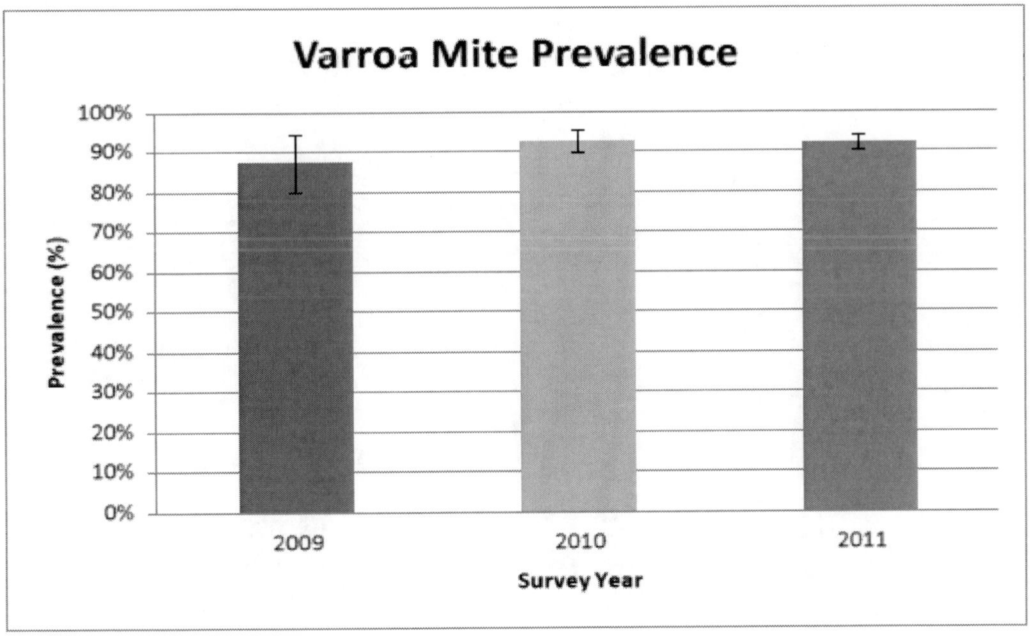

Figure 5. Varroa mite prevalence over 3 years of survey. (95% Confidence Intervals shown).

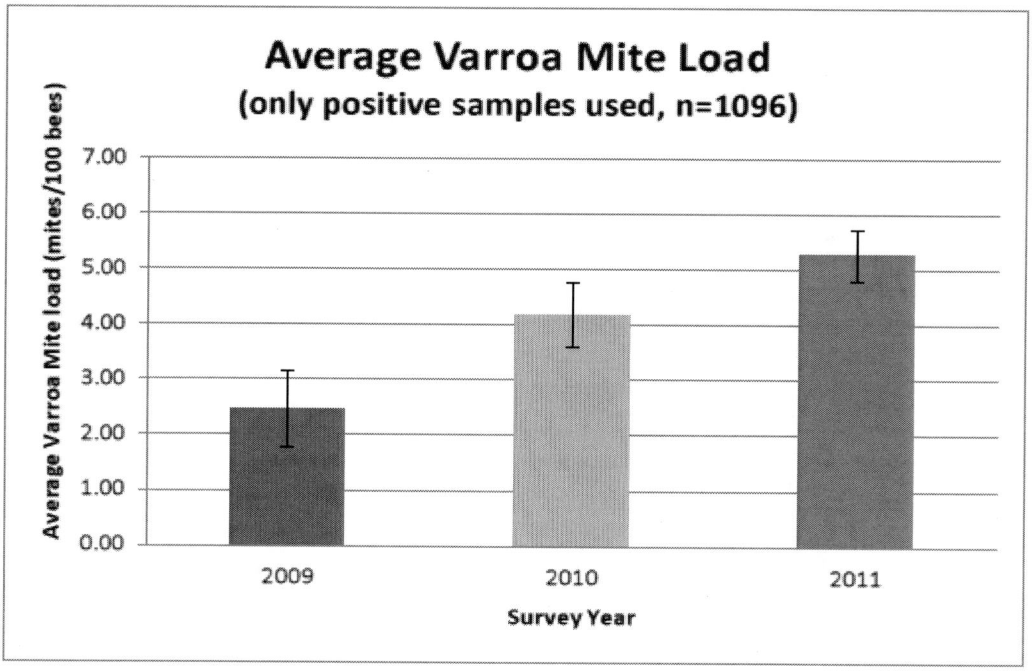

Figure 6. Average Varroa mite load over 3 years of survey. (for samples testing positive) (95% Confidence Intervals shown).

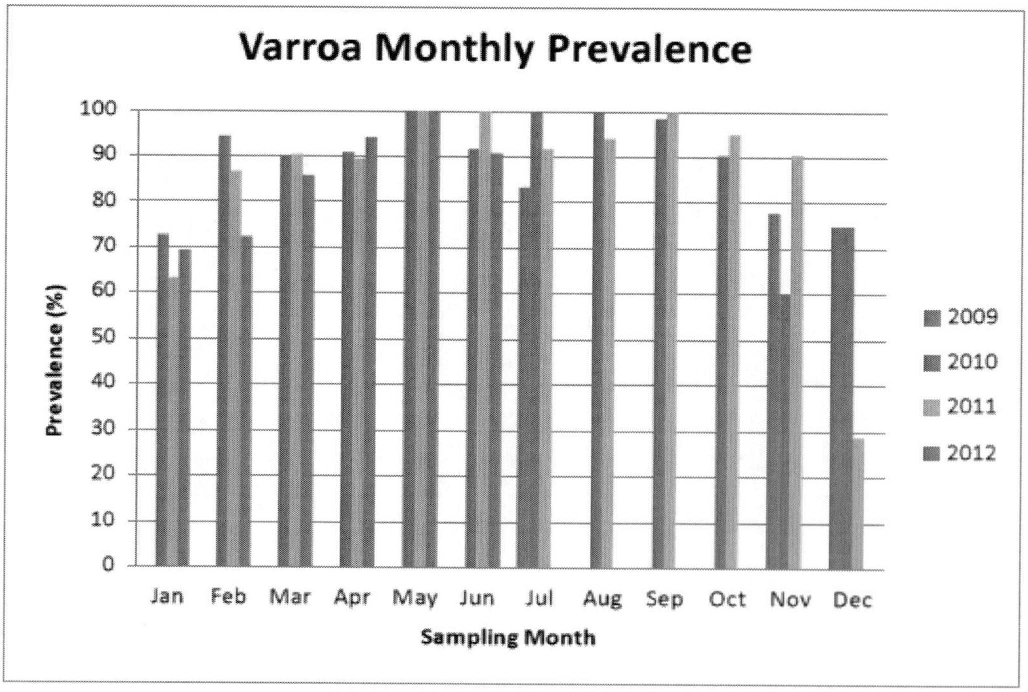

Figure 7. Monthly prevalence for Varroa mites.

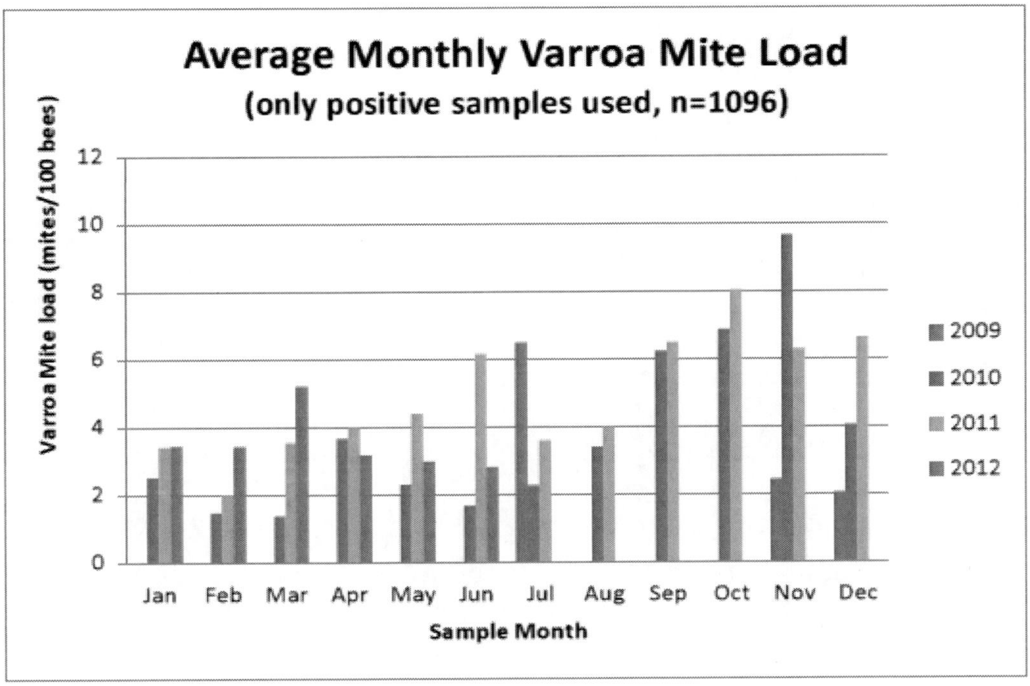

Figure 8. Average monthly Varroa mite load. (for samples testing positive).

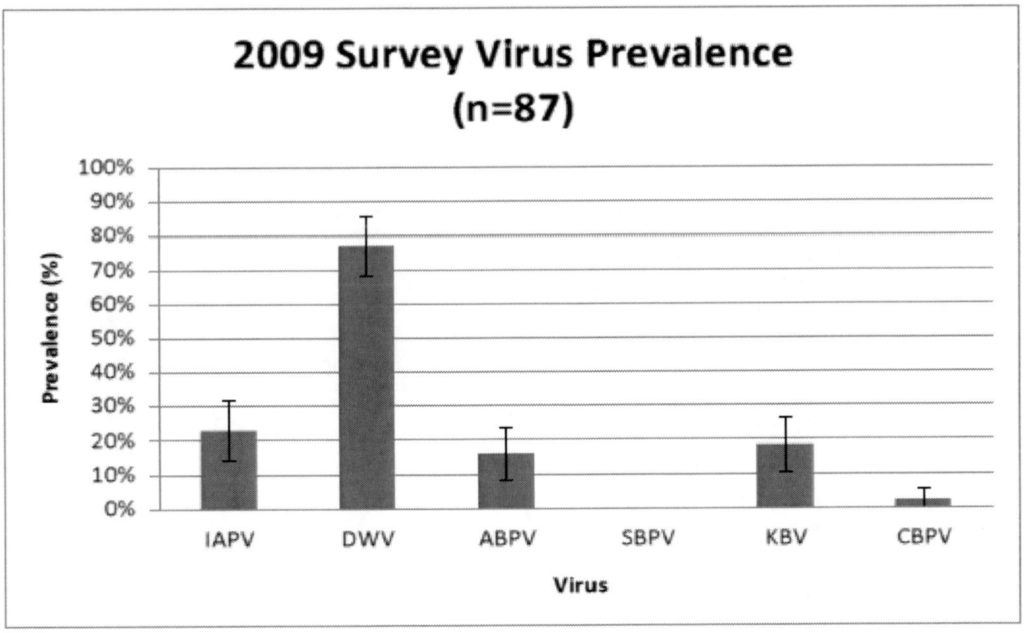

Figure 9. Virus prevalence for 2009 Pilot Survey. (95% Confidence Intervals shown).

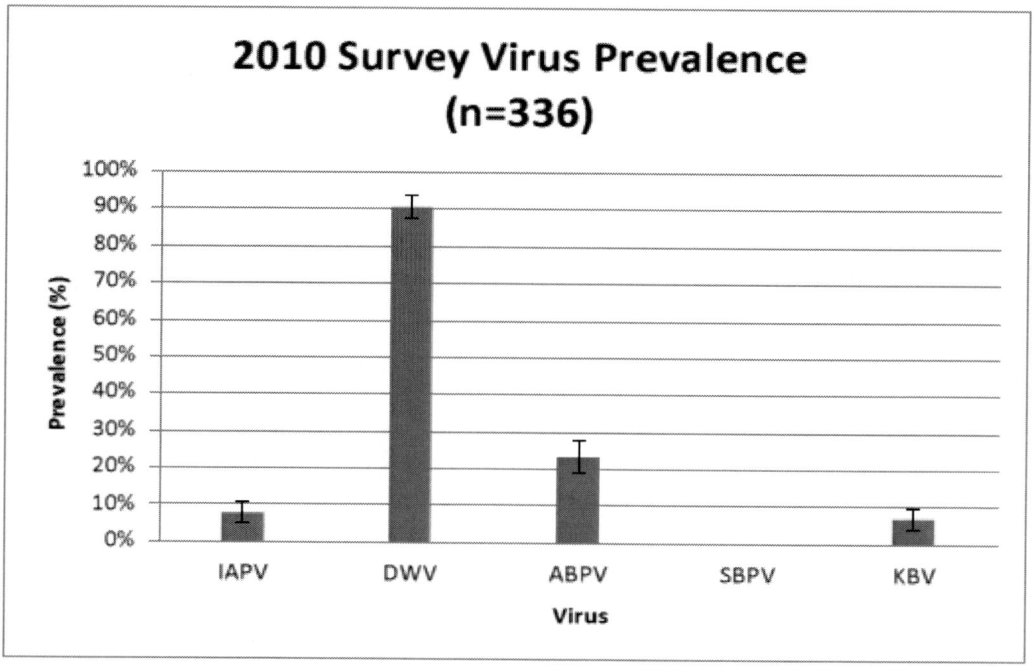

Figure 10. Virus prevalence for 2010 Limited National Survey. (95% Confidence Intervals shown).

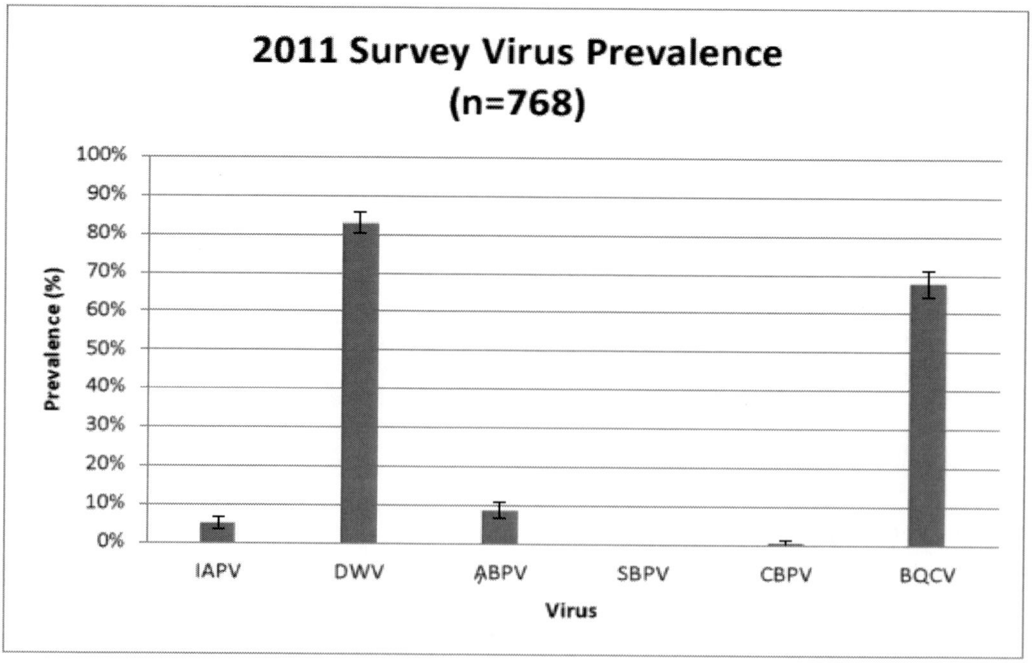

Figure 11. Virus prevalence for 2011 National Survey. (95% Confidence Intervals shown).

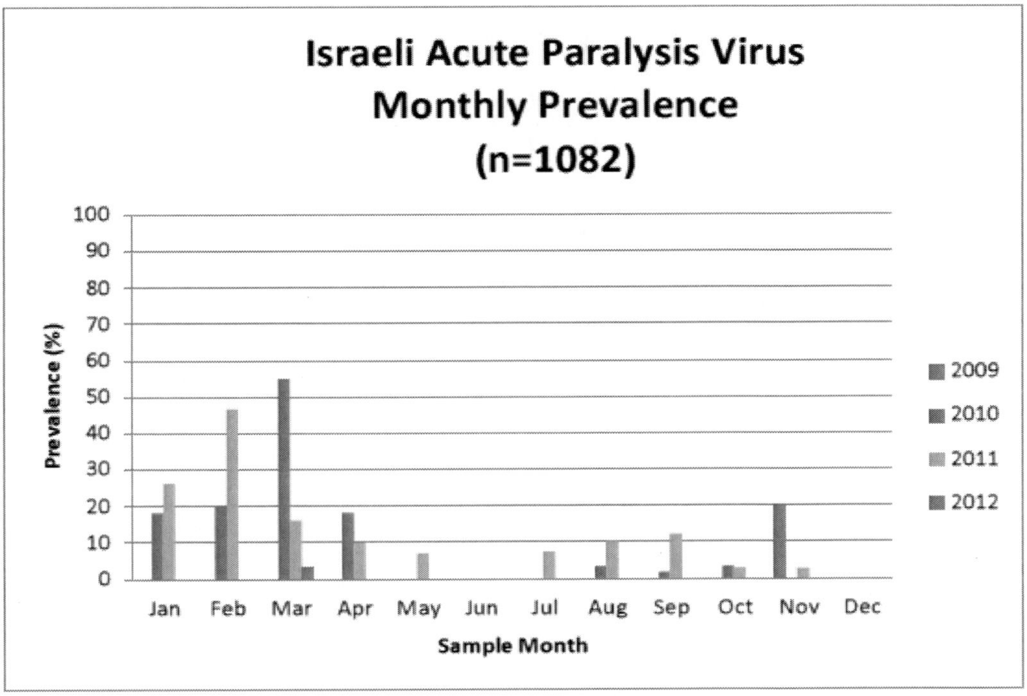

Figure 12. Israeli Acute Paralysis Virus prevalence over 3 years of survey.

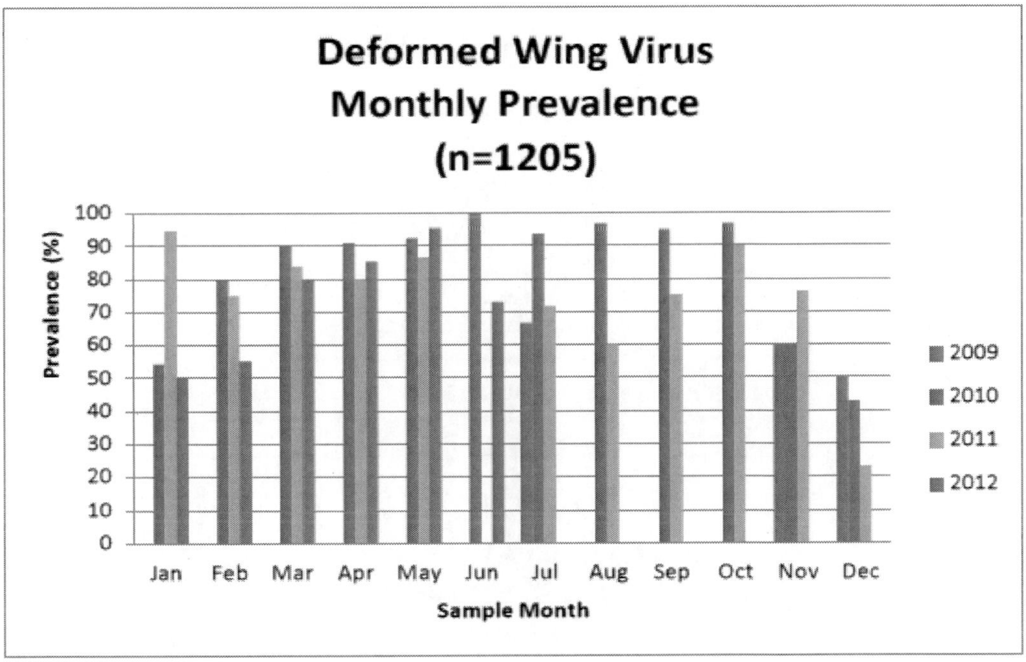

Figure 13. Deformed Wing Virus prevalence over 3 years of survey.

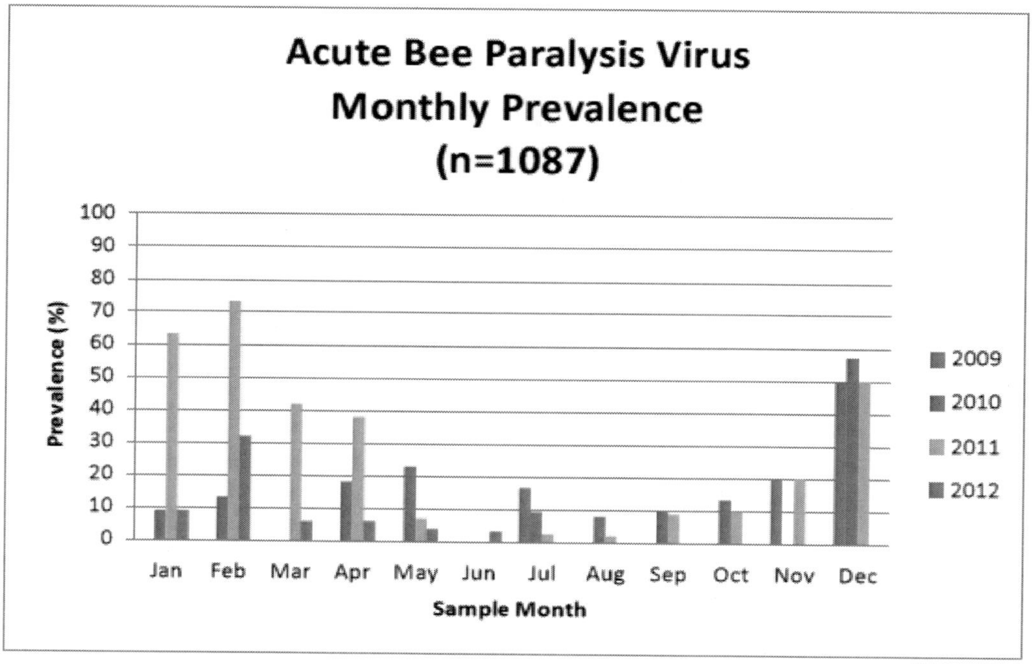

Figure 14. Acute Bee Paralysis Virus prevalence over 3 years of survey.

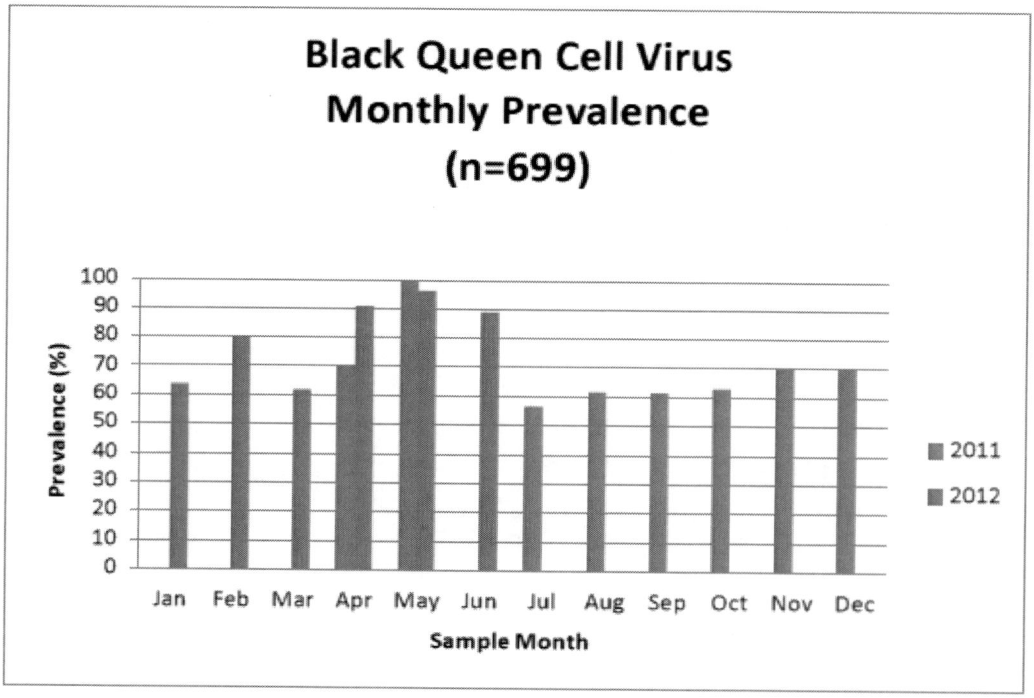

Figure 15. Black Queen Cell Virus prevalence, 2011/2012 only.

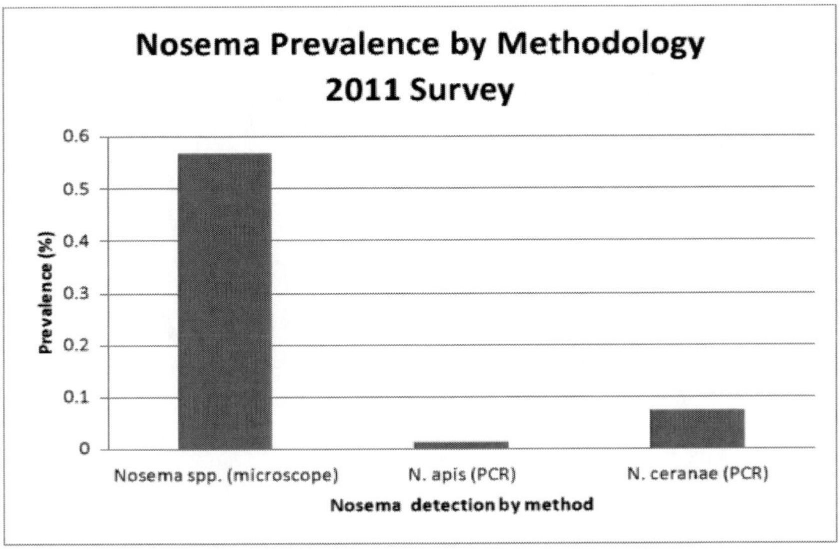

Figure 16. Nosema prevalence by methodology.

Pesticide	LOD (ppb)	Prevalence n=99 (%)	Average detection if positive for target (ppb)	Range (ppb)	Description
Azoxystrobin	2	5.1	13.7	6.4-30.8	Fungicide
Carbendazim (MBC)	5	5.1	105.3	8.1-233	Fungicide
Cyprodinil	4	3.0	19.2	16-22.4	Fungicide
Fenbuconazole	2	2.0	205.0	74.9 - 335	Fungicide
Metalaxyl	2	1.0	37.9	37.9*	Fungicide
Pyraclostrobin	15	1.0	56.5	56.5*	Fungicide
THPI	50	6.1	2360.2	37.6 - 7060	Fungicide
Atrazine	6	1.0	51.3	51.3*	Herbicide
Oxyfluorfen	1	2.0	3.1	1.7-4.4	Herbicide
Pendimethalin	6	5.1	23.2	5.1-43.1	Herbicide
Trifluralin	1	2.0	1.3	1-1.5	Herbicide
2,4 Dimethylphenyl formamide (DMPF)	4	27.3	100.6	10-573	Insecticide
Aldicarb sulfone	3	1.0	1.0	14.0*	Insecticide
Aldicarb sulfoxide	20	1.0	35.9	35.9*	Insecticide
Bifenthrin	1	8.1	4.2	1.2-11.4	Insecticide
Chlorpyrifos	1	20.2	6.5	1.1-21.4	Insecticide
Coumaphos	1	34.3	87.5	2-1110	Insecticide
Coumaphos oxon	1	5.1	14.4	6.2-23.5	Insecticide
Cyfluthrin	4	1.0	3.9	3.9*	Insecticide
Cyhalothrin total	1	7.1	9.5	2.2-36.2	Insecticide
Cypermethrin	4	1.0	9.3	9.3*	Insecticide
Dieldrin	10	1.0	12.4	12.4*	Insecticide
Diflubenzuron	20	1.0	84.3	84.3*	Insecticide
Endosulfan I	2	5.1	38.7	2.2 - 124	Insecticide
Endosulfan II	2	3.0	16.1	2.1-39.5	Insecticide
Endosulfan sulfate	2	3.0	19.6	1.6 - 50.4	Insecticide
Esfenvalerate	2	4.0	5.6	3.7-7.3	Insecticide
Fenpropathrin	1	6.1	43.2	20.7-93.6	Insecticide
Fenpyroximate	5	11.1	28.4	5.5 - 114	Insecticide
Fluvalinate	1	38.4	39.9	2.2-182	Insecticide
Imidacloprid	1	9.1	30.8	3.5-216	Insecticide
Permethrin total	10	1.0	20.0	20.0*	Insecticide
Phosmet	10	1.0	785.0	785.0*	Insecticide
Thiacloprid	1	2.0	187.6	49.1-326	Insecticide
Thymol	50	27.3	2271.8	37.5 - 39700	Insecticide

Figure 17. Pesticide analysis for 2011 survey. (99 samples) (*denotes single detection only).

REFERENCE

Evans J.D. (2006). Beepath: An ordered quantitative-PCR array for exploring honey bee immunity and disease, Journal of Invertebrate Pathology, 93 (2), pp. 135-139.

In: Bee Health: Factors, Analyses, and Research Progress ISBN: 978-1-62618-044-4
Editor: Roger Dunham

Chapter 5

COLONY COLLAPSE DISORDER PROGRESS REPORT, JUNE 2012*

CCD Steering Committee

EXECUTIVE SUMMARY

This report describes research progress on Honey Bee Colony Collapse Disorder during 2011. Mandated by Congress in the 2008 Farm Bill (section 7204[h][4]), this fourth annual report represents the work of a large number of scientists from 8 Federal agencies, 2 State departments of agriculture, 22 universities, and several private research efforts.

In response to unexplained losses of U.S. honey bee colonies that began to be reported in 2006 as a condition named Colony Collapse Disorder (CCD), USDA's Agricultural Research Service (ARS) and National Institute of Food and Agriculture (NIFA) led a collaborative effort to define an approach for characterizing and resolving the problem; this effort resulted in the CCD Action Plan in July 2007. Many universities and Federal, State, and private organizations helped develop the plan and, through 2011, continued to carry out work to address CCD.

A survey of managed honey bee colonies found overall losses of 22 percent on average during winter 2012 (October 2011 to April 2012), down from losses of approximately 33 percent reported in 2010. In general, in the years since CCD began to be reported, winter losses have been averaging around 33 percent, of which approximately one-third was attributed to CCD. Many factors could be involved in the lower level of overall losses in 2011, including the mild winter weather in many parts of the United States. Nevertheless, the overall proportion of winter losses is still high, and the continued economic viability of pollination by honey bees remains threatened.

Although a number of factors continue to be associated with CCD, including parasites and pathogens, poor nutrition, pesticides, bee management practices, habitat fragmentation, and agricultural practices, no single factor or pattern of factors has been proven to be "the cause" of CCD.

---

* This report was released by the USDA Agricultural Research Service, June 2012.

Some recent scientific evidence appears to indicate that some pesticides may have sublethal effects on honey bees; however, the relevancy of some of these effects to estimating overall risks to honey bees is uncertain. In early 2010, ARS held a workshop with the U.S. Environmental Protection Agency to discuss how the potential effects of pesticides could be better documented in laboratory and field-based studies. ARS scientists and others have been examining whether pesticide-related effects may correlate with CCD incidents or other bee health problems.

The CCD Steering Committee is examining the current accumulation of knowledge concerning CCD and has begun the process of identifying new and relevant research needs to determine the cause or causes of CCD and how other factors or specific combinations of factors contribute to declining pollinator health. Scientists, beekeepers, growers, and other stakeholders have been invited to convene in October 2012 to update the CCD Action Plan for the next 5 years (2013–2017). Participants in the October meeting will consider the development of strategies to address the multiple factors (parasites and pathogens, poor nutrition, pesticides, and bee management practices) associated with diminished pollinator health in general, rather than focus exclusively on honey bees and CCD.

Research continued in 2011 in the four topic areas outlined in the 2007 CCD Action Plan. The studies encompassed honey bee and pollinator losses in general, not just CCD. Progress in the four topic areas is highlighted below.

- Topic I: Survey and Sample Data Collection. Surveys continued to provide evidence of high honey bee losses due to a combination of CCD and declines in pollinator health. Research has shown that weak colonies had overall higher pathogen levels and evidence of pesticide residues, although no pattern of specific pathogens or pesticides was indicated. Colonies in comparatively good health also contained a wide range of pesticide residues.
- Topic II: Analysis of Existing Samples. Previously, viruses and other pathogens and parasites were found to be present at greater levels in colonies affected by CCD than in those not affected by CCD. Studies in 2010 revealed several new viruses and other pathogens affecting honey bees. Additional studies to determine whether these new pathogens are involved in CCD-affected hives are continuing.
- Topic III: Research to Identify Factors Affecting Honey Bee Health, Including Attempts to Recreate CCD Symptomology. CCD was initially characterized by the rapid loss of adult worker bees from the colony, lack of dead worker bees in or immediately outside of the hive, and delayed invasion of hive pests. Researchers have observed that the Varroa mite and other pathogens such as Nosema may be contributing factors to CCD, although these occur at levels that are typically considered below economic damage thresholds. Researchers continued their efforts to document whether correlations exist between overall colony health and the presence of Varroa mites, diverse pathogens, and pesticides.
- Topic IV: Mitigation and Management Preventive Measures. Two national multi-year projects, the ARS Area-wide Project on Honey Bee Health and a NIFA-funded Coordinated Agricultural Project, progressed in developing management strategies to combat bee losses. The "Bee Informed Partnership" (http://beeinformed.org), sponsored by NIFA, continued to examine bee management practices and facilitate

communication among beekeepers of practices that reduce colony losses while promoting bee health. The eXtension Community of Practice (www.extension.org/bee health) also continued to disseminate information on honey bee health and management practices. A National stakeholders' meeting scheduled for October 15–17, 2012, is intended to identify management practices for the different factors affecting honey bee health.

The Natural Resources Conservation Service has revised its appropriate conservation practices to encourage landowners to provide quality pollinator habitat and protect pollinators.

INTRODUCTION

This is the fourth annual report prepared in response to a requirement in the 2008 Farm Bill, section 7204(h)(4), which directed the Secretary of Agriculture to—

> "submit to the Committee on Agriculture of the House of Representatives and the Committee on Agriculture, Nutrition, and Forestry of the Senate an annual report describing the progress made by the Department of Agriculture in:
> a) Investigating the cause or causes of honey bee colony collapse; and
> b) Finding appropriate strategies to reduce colony loss."

BACKGROUND AND HIGHLIGHTS OF RESEARCH

After the large-scale, unexplained losses of managed honey bee (*Apis mellifera* L.) colonies began in the United States during the winter of 2006–2007, researchers identified a set of symptoms that were termed "Colony Collapse Disorder" (CCD). In response to these continued losses, Federal and State government, university, and private researchers, led by the Department of Agriculture's (USDA) Agricultural Research Service (ARS) and National Institute of Food and Agriculture (NIFA), mobilized to define an approach to CCD, an effort that resulted in the formation of the CCD Steering Committee and publication of the CCD Action Plan in July 2007. As of 2011, the CCD Steering Committee included representatives from various USDA offices, including ARS, NIFA, the Animal and Plant Health Inspection Service (APHIS), the Natural Resources Conservation Service, the National Agricultural Statistics Service (NASS), and the Office of Pest Management Policy, and the Environmental Protection Agency's (EPA) Office of Pesticide Programs. In addition, many public and private organizations are actively working to remedy CCD.

CCD is a complex syndrome that has been hard to define and combat. Thus, developing effective solutions to the problem requires a considerable commitment of Federal resources.

During the past 5 years (2006–2011), numerous causes for CCD have been proposed and examined. Many potentially associated factors have been identified throughout the course of research, a number of which appear to have a high correlation with the pattern of CCD incidents. However, the strength of these associations has varied considerably, and it has become increasingly clear that no single factor is responsible for the syndrome. In addition, research has not been able to determine whether all cases of CCD are caused by the same set of factors or the same factors in a particular combination.

Independent studies have shown that bees are exposed to a wide range of pesticides. Pesticides found in colonies range from those used to control pathogens or pests that adversely affect honey bees, to commercial agricultural products. A survey of bees, honey, and comb for the presence of 170 pesticides or pesticide residues performed in 2010 did not find any pattern of exposure that correlated with CCD incidents, which would be expected if pesticides were a major factor in causing CCD. The pesticides detected with the greatest frequency and in the largest quantities were those used by beekeepers to control *Varroa* mites. Pesticide effects on bees continue to be a subject area of intense research.

Other data indicate that some pesticides at high concentrations interact with other pesticides, honey bee parasites, or pathogens in ways that significantly increase individual bee mortality rates. Further studies are needed to ascertain whether these synergistic effects occur at environmentally relevant concentrations or whether managed honey bee colonies are commonly exposed to these levels of pesticides.

In addition, some studies have shown that exposure to certain pesticides can have sublethal deleterious effects on bee behavior. However, while individual bees were found to be adversely affected by sublethal exposure to some pesticides under experimental conditions, further studies are needed to show whether a colony's abilities to pollinate crops, produce honey, and maintain overall population are compromised by a specific sublethal effect of pesticide exposure on individual bees.

In an effort to address the multiple factors associated with pollinator declines, the CCD Action Plan was organized under four areas: (1) survey and sample data collection; (2) analysis of existing samples; (3) research to identify factors affecting honey bee health, including attempts to recreate CCD symptomology; and (4) mitigation and preventive measures. Summaries of research under each of the four topic areas are presented below.

Topic I: Survey and Sample Data Collection

A survey of beekeepers throughout the United States was jointly conducted for the fifth consecutive year by the Apiary Inspectors of America and ARS with additional assistance this year from the Bee Informed Partnership.[1] Total average losses of honey bee colonies for winter 2011–2012 (October 2011–April 2012) were 22 percent, which is lower than previous surveys performed between 2007 and 2010.[2] CCD is characterized by a sudden dwindling of bees as winter turns to spring with a queen, some brood, and a few nurse bees remaining in the hive, but at levels below those needed (10,000 bees) for a colony to survive as it begins to forage and rear brood. The unusually warm 2011–2012 winter could be one contributing factor to the drop in colony losses, although no direct scientific research has been performed to discern whether a connection exists between weather and CCD. The National Oceanic and Atmospheric Administration reported that January 2012 was the fourth warmest January on record.

Of beekeepers responding to the survey who reported losing any colonies, 37 percent said they lost at least some of their colonies with no dead bees present, which is indicative of CCD (http://beeinformed.org/2012/05/ winter2012/).

APHIS continued its survey of beekeepers in 34 States to detect exotic pests and diseases of honey bees. The survey conducted thus far has not detected *Apis ceranae,* the Asiatic honey bee; the slow paralysis virus, which has been reported in Australia; or the parasitic mite *Tropilaelaps,* which is commonly found in Asia on several species of honey bees; so these pests have likely not invaded the United States. *Nosema ceranae,* a microsporidial pathogen that was recently introduced into the United States, was the dominant species of *Nosema* detected. *N. ceranae* has been tentatively linked in some studies as contributing to CCD in the United States, although no conclusive evidence has been found.

Topic II: Analysis of Existing Samples

A series of colony samples collected from throughout the United States in 2006 and 2007 was further analyzed using DNA sequencing in an effort to detect the presence of exotic or novel pathogens that might have been overlooked in studies that were conducted in 2007.[3]

Analysis of a broad set of healthy and CCD-affected hives did not reveal a sole causative agent but rather a host of viral, bacterial, and fungal pathogens that occurred at higher levels in CCD-affected hives.[4]

Recent efforts by Cornman and colleagues[4] did not identify a single unique pathogen that could be a CCD causative agent, but the data did lend support to earlier studies suggesting that whereas a complex set of pathogens may be involved in the cause of colony losses, no single pathogen can be solely linked to CCD.

Topic III: Research to Identify Factors Affecting Honey Bee Health, Including Attempts to Recreate CCD Symptomology

Research efforts jointly supported by ARS and NIFA continue to investigate numerous factors alone or in combination that may play a role in causing CCD. These include parasites and pathogens, pesticides, poor nutrition, bee management practices, and to a lesser extent, other pests such as the small hive beetle (*Aethina tumida*).

The *Varroa destructor* mite remains one of the primary threats to honey bee health. Several new and existing *Varroa* mite control agents that are being tested by ARS researchers may help to control these mites:

- A new chemical product is in the final year of testing for overall efficacy. This product holds great promise but is still proprietary and unavailable at this time.
- A new product, Apivar (Véto-pharma, Villebon-sur-Yvette, France), which contains the active ingredient amitraz (a nonsystemic acaricide and insecticide), has been tested by ARS researchers. Apivar was found to have good efficacy, and there were no observed adverse bee health issues. The ARS laboratory in Beltsville, Maryland, is proposing residue testing for approval by the EPA to support efforts to register the pesticide in the United States for use during the winter (i.e., not during honey flow, which is that period in the year when one or more major sources of nectar are in bloom).

- A relatively new product, HopGuard®,5 which uses beta plant acids from hops plants, has been formulated and is now in commercial production. There is no evidence to suggest that the product harms bees or broods, and it leaves no residue. It is highly effective in reducing mite populations when applied to colonies without capped brood (covered brood cells in which bee larvae and pupae are present), or in package bees (a screen box that includes a queen and worker bees). Because multiple applications are needed, HopGuard is being used primarily by hobbyists who have more time than commercial beekeepers to attend to colonies; however, research is underway to develop a slow-release formulation acceptable to all operations.

Pesticide Effects

Field exposure of pollinator insects, including honey bees, to pesticides at both lethal and sublethal levels continues to be a concern. Extensive discussions have occurred in the media and among researchers about whether pesticides are a significant threat to pollinators. In particular, exposure to pesticide-contaminated dust from abrasion of certain pesticide-treated seed (e.g., corn) during spring plantings appears to have negative effects on individual honey bees in experimental (laboratory and field) settings.

Dust collected from within mechanized planters during spring planting has been shown to contain pesticide concentrations in parts per thousand (i.e., concentrations 1 million times greater than parts per billion).[6]

However, there are no data to indicate that bees are exposed to residue levels in the environment that even approach those measured in the seeding equipment. Researchers have not yet determined how honey bees and other pollinators react when exposed to environmentally relevant levels and whether a significant number of bees would likely be present in or around fields that are planted or seeded. Also, recent advances in seeders are reducing the hazard associated with pesticide-contaminated dust.

In study results published in 2012,[7] exposure of individual, immature honey bees to sublethal levels of neonicotinoids resulted in increased susceptibility to the gut pathogen *Nosema,* although the response was not concentration dependent.

The colony's overall health, population, ability to gather nectar, and pollinate were not affected by these sublethal effects on individual bees, but this study does demonstrate that there are complex interactions among various factors that could contribute to weakening individual bees and making them more susceptible to additional perturbations.

When pesticides are viewed in aggregate on a national scale, residues of pyrethroids (a large class of man-made pesticides similar to the natural pesticide pyrethrum produced from Chrysanthemum flowers) pose a threefold greater hazard to bee colonies than neonicotinoids, based on mean and frequency of detection in pollen samples and relative acute toxicity.[8] The synthetic pyrethroid detected in the highest quantity and frequency in honey bee colonies that is used by beekeepers to control *Varroa* mite is tau fluvalinate.[9] Because pyrethroids are nonsystemic, adoption of agricultural practices that mitigate exposure should be pursued. Additionally, when honey bees were challenged with both pesticides and *Varroa* mites, they showed increased immune response.[10]

Additional studies—on diet and pathogen interactions—revealed that nutrition could modulate viral infections.[11]

Nutrition and Nosema Effects

There were significant efforts in 2011 to determine the effects and outcomes of feeding honey bee colonies commercial diets and supplements, especially as a support for overwintering colonies. Many of the commercial diets were determined sufficient to increase brood production and adult bee populations.

A new pollen substitute (MegaBee®, S.A.F.E. R&D, distributed by Dadant & Sons, Inc., Hamilton, Ill.) is now in commercial production. When MegaBee® is fed in liquid or solid form, it stimulates brood production and immune response at levels comparable to those when honey bees are fed pollen alone.

Stakeholders have called for a study of the epidemiology of *N. ceranae,* the establishment of economic damage thresholds, and the testing of new products for efficacy in treating *Nosema.* At this point, only a single product, the antibiotic fumagillin,[12] is commercially available to control *Nosema.* Some beekeepers use it as a prophylactic for *Nosema* infections when treatment is not warranted.

To date, no treatment thresholds have been established for the use of the antibiotic, and new guidance on its use is needed to prevent target species resistance that can develop with the continued use of a single product. Several additional products for *Nosema* control were tested by ARS scientists in Weslaco, Texas, but none have yet been found to be effective.

Land Use/Pollinator Effects

Honey bees are essential pollinators for many crops, but in light of CCD and other new threats that have arisen in the last 15 years, researchers have begun assessing the use of alternative insects such as bumble bees (Bombus sp.) and blue orchard bees (Osmia lignaria) to serve as crop pollinators. Unfortunately, even as this research is occurring, new problems in insect management have arisen. For example, a recent study demonstrated a high degree (73–93 percent) of cross-infection of viruses between honey bees and native bumble bees.[13]

Thus, it is possible that co-infections and reinfection pathways may develop, thereby complicating the use of alternative pollinators.

Topic IV: Mitigation and Management Measures

Research to resolve CCD and improve the overall health of pollinator insects is being undertaken primarily through the ARS Area-wide Project on Honey Bee Health and the NIFA Coordinated Agricultural Project (CAP). Funding from ARS and NIFA, with additional contributions by the National Honey Board, the Almond Board of California, Burt's Bees, Häagen-Dazs, the North American Pollinator Protection Campaign, Project Apis m. (PAm), and the Foundation for the Preservation of Honey Bees has supported a variety of new studies. Results from these research efforts continue to be published.

The eXtension[14] Web site (http://www.extension.org/bee%20health), established in 2010, is providing reliable, research-based information to beekeepers and the general public, among other resources.

A new NIFA-funded CAP, the Bee Informed Partnership (http:// beeinformed.org), is endeavoring to decrease the number of managed honey bee colonies that die each winter by helping determine which management practices are working best to keep bees alive. It will also examine various data collections to assess pest and disease levels in an effort to bridge the gap between research and extension and present the best management information. In addition, the Bee Informed Partnership has joined ARS and the Apiary Inspectors of America in carrying out the annual survey of winter losses of managed honey bee colonies that have occurred since 2007.

A Bee Team has been established at the University of California Cooperative Extension office in Oroville, California, with funding from the CAP program, the University of Minnesota, National Honey Board, and the Almond Board of California. The goal of the team is to help beekeepers monitor diseases and arthropod pests and select for more disease- and mite-resistant breeder stocks. The Bee Team will test for hygienic behavior (i.e., behavior in which mite-infested brood are removed from the hive), the presence of *Varroa* mites, and *Nosema* spp. in at least 50 colonies at each participating bee breeder's operation 3 times each year. Data from the samplings will be provided to each bee breeder to help them make informed decisions on choosing breeder queens and appropriate treatments. The bee breeders have agreed to pay a fee for these services so that in the future, the Bee Team can be self sustaining. If this model is determined to work well, the Bee Team hopes to establish similar teams to assist queen producers in the Southeast and other parts of the United States.

Finally, another NIFA-funded CAP program has published a best management practices guide, educational videos, and health bulletins for beekeepers. These materials are also available through the Bee Health Community Page (http://www.extension.org/bee_health), which serves as a repository of peer-reviewed, credible scientific information for the bee community.

In an effort to provide healthy habitat for all pollinators, the Natural Resources Conservation Service has revised appropriate conservation practices to encourage landowners to include in their conservation seed mixes and practices vegetation that will provide desired forage for pollinators from early spring to late fall and protect pollinators.

SUMMARY

While research has developed new scientific information about CCD and honey bee health and management in general during the past 5 years, the complex problems of maintaining a strong pollinator industry have only grown more complicated. What is clear is that researchers must look beyond simple one-factor causes of bee decline and losses. This research is multifactoral and more challenging. In light of the complexities involved in pollinator health, research continues to seek scientific facts on bee health and ultimately to safeguard the health of the country's critical pollinators.

End Notes

[1] Bee Informed, sponsored by NIFA, is an extension project that is trying to decrease the winter mortality of managed honey bee colonies.

[2] Average losses were 31 percent during the 2007 winter, 35 percent in 2008, 29 percent in 2009, 34 percent in 2010, and 30 percent in 2011.

[3] Cox-Foster DL, Conlan S, Holmes EC, et al. Metagenomic survey of microbes in honey bee Colony Collapse Disorder. Science. 2007:318;283.

[4] Cornman RS, Tarpy DR, Chen Y-P, Jeffreys L, Lopez D, Pettis JS, vanEngelsdorp D, Evans JD. Pathogen webs in collapsing honey bee colonies. PloS One. (In press.)

[5] HopGuard is produced by BetaTec Hop Products (Washington, D.C.) under a Cooperative Research and Development Agreement with ARS.

[6] Krupke CH, Hunt GJ, Eitzer BD, Andino G, Given K. Multiple routes of pesticide exposure for honey bees living near agricultural fields. PLoS ONE. 2012;7(1): e29268. DOI: 10.1371/journal.pone.0029268.

[7] Pettis J, vanEngelsdorp D, Johnson J, Dively G. Pesticide exposure in honey bees results in increased levels of the gut pathogen Nosema. Naturwissenschaften. 2012;99:153–158.

[8] Frazier JL, Frazier MT, Mullin CA, Ashcraft S. Pesticides and their involvement in Colony Collapse Disorder. Am Bee J. 2011;August:779–784.

[9] Mullin C, Frazier M, Frazier J, Ashcraft S, Simonds R, vanEnglesdorp D, Pettis J. High levels of miticides and agrochemicals in North American apiaries: implications for Honey Bee Health. Plos One. 2010;5(3):39754.

[10] Gregorc A, Evans JD, Scharf M, Ellis JD. Gene expression in honey bee (Apis mellifera) larvae exposed to pesticides and Varroa mites (Varroa destructor). J Insect Physiol. 2012;April 9 [Epub ahead of print].

[11] DeGrandi-Hoffman G, Chen Y, Huang E, Huang M. The effect of diet on protein concentration, hypopharyngeal gland development and virus load in worker honey bees (Apis mellifera L.) J Insect Physiol. 2010;56:1184–1191.

[12] Fumagillin, a complex biological molecule, was isolated in 1949 from the bacterium Aspergillus fumigates. It is used today as an antimicrobial agent.

[13] Singh R, Levitt AL, Rajotte EG, et al. RNA viruses in Hymenopteran pollinators: Evidence of inter-taxa virus transmission via pollen and potential impact on non-Apis Hymenopteran species. PLoS ONE. 2010;5(12): e14357. DOI: 10.1371/journal.pone.0014357.

[14] The eXtension initiative is an internet-based educational partnership of 74 Land Grant universities in the United States that operates under the auspices of the eXtension Foundation, which broadly functions as part of NIFA's Cooperative Extension System.

In: Bee Health: Factors, Analyses, and Research Progress ISBN: 978-1-62618-044-4
Editor: Roger Dunham

Chapter 6

COLONY COLLAPSE DISORDER PROGRESS REPORT, JUNE 2011*

CCD Steering Committee

EXECUTIVE SUMMARY

Mandated by section 7204(h)(4) of the Food Conservation and Energy Act, Pub. Law No. 110-246, the "2008 Farm Bill," this third annual report of research progress during 2010 on Honey Bee Colony Collapse Disorder (CCD) represents the work of a large number of scientists from 8 Federal agencies, 2 State departments of agriculture, 22 universities, and several private research efforts.

In response to the unexplained losses of U.S. honey bee colonies now known as Colony Collapse Disorder (CCD), USDA's Agricultural Research Service (ARS) and National Institute of Food and Agriculture (NIFA) led a collaborative effort to define an approach for responding to CCD, resulting in the CCD Action Plan, in July 2007. Other Federal agencies e.g., the Animal and Plant Health Inspection Service (APHIS), the Natural Resources Conservation Service, and the Environmental Protection Agency's (EPA's) Office of Pesticide Programs have been members of the CCD Steering Committee for many years; this year, USDA's Office of Pest Management Policy and the National Agricultural Statistics Service have joined the Steering Committee. Many universities and organizations (i.e., Federal, State, and private) were also involved in developing this plan and are carrying out work that addresses the CCD problem.

Annual surveys clearly show that overall colony losses continue to be as high as 30 percent or more since CCD began to be reported. Beekeepers cannot economically sustain such high levels of losses indefinitely. Although a number of factors have been associated with CCD and pollinator declines in general, no single factor or specific combination of factors has been identified as a "cause." Factors associated with declines include disease/parasites, nutrition, pesticides, bee management practices, habitat fragmentation, and

* This is an edited, reformatted and augmented version of report released by the USDA Agricultural Research Service, dated June 2011.

agricultural practices. Reducing the incidence of CCD and pollinator declines will likely require managing multiple factors simultaneously. Overall losses to managed pollinators were about 33 percent. Relative to the overall losses, CCD contributed approximately 8 percent in recent national surveys. However, losses to individual beekeepers attributed to CCD may vary significantly by size of the beekeeping operation. Therefore, this statistic is just a rough estimate.

To help ensure more robust pollinator health, the CCD Steering Committee is facilitating the development of science-based best management strategies that integrate current research results associated with CCD incidence and pollinator decline in general. As part of these efforts, the CCD Steering Committee member agencies are working together to examine ways to mitigate impacts on pollinator health. This effort requires coordination across different Federal partners (USDA, Department of Interior, and EPA), States, and stakeholders to ensure a sustained effort that effectively integrates management options.

The CCD Steering Committee believes that it is necessary for scientists, beekeepers, and growers to reconvene in the next year to revisit approaches and research, which are now needed to counter honey bee decline, taking into account information developed during the past 3 years and to revise the 2007 Action Plan to reflect the current state of knowledge. A range of stakeholders will be engaged to develop strategies for managing factors associated with diminished pollinator health and not focus exclusively on CCD.

Research is ongoing in the four topic areas outlined in the CCD Action Plan, but the studies encompass factors involved in bee losses in general, not just due to CCD. The four areas are **(1) survey and sample data collection, (2) analysis of existing samples, (3) research to identify factors affecting honey bee health, including attempts to recreate CCD symptomology, and (4) mitigation and management preventive measures.** Progress in each of the four topic areas is highlighted below.

- ***Topic I: Survey and Sample Data Collection.*** Surveys continue to provide evidence of high honey bee losses due to CCD and declines in pollinator health. Correlated with these high levels of losses, research has shown that weak colonies had overall increased pathogen levels and showed evidence of pesticide residues, although no pattern of specific pathogens or pesticides was indicated. Colonies in comparatively good health also contained a wide range of pesticide residues. To understand the potential role of pesticides in pollinator declines, representatives of the CCD Steering Committee (APHIS, ARS, and EPA) participated in a Society of Environmental Toxicology and Chemistry global Pellston conference. The purpose was to develop a risk assessment process for honey bees and non-*Apis* bees and to identify the exposure and effects data needed to inform that process.
- ***Topic II: Analysis of Existing Samples.*** Previously, viruses and other pathogens and parasites were found to be present at greater levels in CCD colonies than in non-CCD colonies. The extent to which pesticides are associated with CCD remains uncertain, and additional research is necessary. Studies in 2010 revealed several new viruses and other pathogens affecting honey bees. Further studies are needed to determine if these new pathogens are involved in CCD-affected hives.
- ***Topic III: Research to Identify Factors Affecting Honey Bee Health, Including Attempts to Recreate CCD Symptomology.*** CCD was initially characterized by the rapid loss of adult worker bees from the colony, lack of dead worker bees, and

delayed invasion of hive pests. Although additional studies are needed, researchers have recently observed that *Varroa* mite and other pathogens such as *Nosema* may be contributing factors to CCD. In addition, the *Varroa* mite and other pathogens occur at levels that are typically considered below economic thresholds. Researchers continue their efforts to document whether there are correlations between the presence of *Varroa* mites, diverse pathogens, and pesticides, which appear to impact overall colony health. In addition, apiaries surrounded by intensively farmed landscapes were found to be associated with higher colony losses in the spring.

- ***Topic IV: Mitigation and Management Preventive Measures.*** Two national multiyear projects, the ARS Area-wide Project on Honey Bee Health and a NIFA-funded Coordinated Agricultural Project (CAP), are continuing to make progress in developing managements strategies to combat bee losses. A new CAP project funded by NIFA, the "Bee Informed Partnership" (http://beeinformed.org), has begun to examine bee management practices and facilitate communication of successful practices between beekeepers. The eXtension Community of Practice (www.extension.org/bee_health) is also disseminating information on honey bee health and management practices.

Colony Collapse Disorder (CCD) Annual Progress Report

This report is the third annual report prepared in response to section 7204(h)(4) of the 2008 Farm Bill, which directed the Secretary of Agriculture to—

> "submit to the Committee on Agriculture of the House of Representatives and the Committee on Agriculture, Nutrition, and Forestry of the Senate an annual report describing the progress made by the Department of Agriculture in—
>
> (A) Investigating the cause or causes of honey bee colony collapse; and
> (B) Finding appropriate strategies to reduce colony loss.

Background and Highlights of Research

After the large-scale, unexplained losses of managed U.S. honey bee (*Apis mellifera* L.) colonies during the winter of 2006–2007, investigators identified a set of symptoms that were termed "Colony Collapse Disorder" (CCD). In response to this problem, Federal and State government, university, and private researchers, led by USDA's Agricultural Research Service (ARS) and National Institute of Food and Agriculture (NIFA), mobilized to define an approach to CCD, an effort resulting in formation of the CCD Steering Committee and publication of the CCD Action Plan in July 2007. Over the past several years, the CCD Steering Committee has included representatives from USDA (Animal and Plant Health Inspection Service (APHIS), NIFA, ARS, and Natural Resources Conservation Service [NRCS]) and the Environmental Protection Agency's (EPA) Office of Pesticide Programs. This year, USDA's Office of Pest Management Policy (OPMP) and National Agricultural

Statistics Service (NASS) joined the Steering Committee. Many organizations, public and private, in addition to those represented on the CCD Steering Committee, are involved in the work to address the CCD problem.

During the past 3 years, numerous causes for CCD have been proposed and examined. There have been many associations identified throughout the course of research; however, the strength of these associations has varied considerably, and it has become increasingly clear that no single factor alone is responsible for the malady.

Researchers continue to document elevated pathogen levels in CCD-affected bees, although no single pathogen or group of pathogens has been definitively linked to CCD. In addition, studies that examined colonies for the presence of known honey bee parasites, such as the *Varroa* mite (*Varroa destructor*), tracheal mite (*Acarapis woodi*), and the fungal gut parasite *Nosema* spp., have not found that any of these parasites by themselves occurred at sufficient levels or pattern to explain CCD.

During the past year, several independent studies have shown that bees are exposed to a wide range of pesticides and that some of these pesticides at high concentrations interact with other pesticides, honey bee parasites, or viruses in ways that significantly increase individual bee mortality. Further studies are needed to ascertain whether these greater-than-additive effects occur at environmentally relevant pesticide concentrations. In addition, studies have shown that exposure to pesticides can result in effects on bee behavior. However, further studies on individual bees are needed to better understand colony-level effects. The pesticides detected with the greatest frequency and quantities are those used by beekeepers to control mites.

Conversely, bees exposed to the pesticide coumaphos, which is used by beekeepers to treat *Varroa* mites, appear to have lower levels of CCD, suggesting that reducing infestation of *Varroa* mites could directly or indirectly reduce incidence of CCD. Results from other longterm monitoring projects also suggest that the fungicide chlorothalonil has been associated with a newly identified condition in the hive named "entombed pollen," which is associated with premature bee mortality. However, it is not yet known whether the presence of chlorothalonil within honeybee comb is exclusively associated with entombed pollen. Other factors not yet determined may also have a role. Taken together, these studies illustrate support for the hypothesis that CCD is a result of many different factors, factors that may work independently or in combination.

Because CCD is a complex phenomenon, developing effective solutions to the problem will depend on considerable research commitment. The current coordinated research response is dedicated to resolving the CCD issue, as well as improving overall pollinator health. NIFA's Coordinated Agricultural Project (CAP) and ARS' Area-wide Project on Honey Bee Health are ongoing efforts that have taken broad, regional approaches to solving pollinator losses. Funding from ARS and NIFA, with additional contributions by a number of other sources, including the National Honey Board, the Almond Board of California, Burt's Bees, Haagen-Dazs, the North American Pollinator Protection Campaign, Project Apis m. (PAm), and the Foundation for the Preservation of Honey Bees, has supported a variety of new studies and attracted new expertise to bee health issues. Results from these research efforts are being published, and a new eXtension bee-health Web site (www.extension.org/bee_health) has been assembled to provide reliable research-based information to beekeepers and the general public.

In an effort to address the multiple factors associated with pollinator declines, the CCD Action Plan is organized under four topic areas: (1) Survey and sample data collection, (2) analysis of existing samples, (3) research to identify factors affecting honey bee health, including attempts to recreate CCD symptomology, and (4) mitigation and management preventive measures. Summaries of research under each of the four topic areas are presented below. More detailed accounts of progress on each of the research projects can be found in Appendix 1.

Topic I: Survey and Sample Data Collection

Surveys of beekeepers throughout the United States were jointly conducted for the fourth consecutive year by the Apiary Inspectors of America and ARS. Total losses (not limited to CCD) for winter 2010–2011 were 30 percent, which was in the same range as previous surveys performed between 2007 and 2009. Most beekeepers indicated that this level of loss was economically unsustainable for beekeeping operations. Large-scale commercial beekeepers indicated that losses were due to several contributing factors, including poor queens, *Varroa* mite, pesticides, and CCD.

Preliminary results from systematic surveys of experimental apiaries in seven States managed by NIFA-funded CAP researchers reported that the leading causes of overall losses included intensive agricultural use of the surrounding landscape, pesticide exposure, and *Varroa* mite and diseases. Surveys conducted in Canada and Europe showed that *Varroa* mite and pathogens were consistently the leading factors correlated with declines.

USDA's APHIS expanded its survey of beekeepers from 13 to 34 States to detect exotic pests and diseases of honey bees. The survey conducted thus far has not detected *Apis ceranae,* the Slow Paralysis Virus that has been reported in Australia, or the parasitic mite *Tropilaelaps,* which commonly is found in Asia on several species of honey bees; so these pests have likely not invaded the United States. *Nosema ceranae,* which is a microsporidial pathogen recently introduced into the United States, was the only species of *Nosema* detected in this recent APHIS survey. This species has been tentatively linked in some studies as contributing to CCD in the United States.

Members of the CCD steering committee (EPA and USDA's APHIS and ARS) participated in a global Society of Environmental Toxicology and Chemistry (SETAC) Pellston[1] workshop in January 2011. The global Pellston conference is an important beginning in developing improved approaches for characterizing potential sublethal impact of pesticides on individual bees and on colonies, as well as to identify the data needed to inform the risk assessment process. The SETAC conference also examined the adequacy of tests performed with *Apis mellifera* to serve as a model to predict effects on other "non-*Apis*" pollinators.

Topic II: Analysis of Existing Samples

As reported in the 2010 Progress Report www.ars.usda.gov/is/br /ccd/ ccdprogressreport2010.pdf, a large survey revealed that levels of pesticides in wax and pollen were similar in both CCD and healthy colonies. Further analysis showed that levels of the varroacide coumaphos were higher in healthy control colonies relative to CCD colonies. These preliminary data suggest several possibilities that need to be examined with more detailed, hypothesis-driven experiments. Molecular studies by scientists at the University of Illinois seem to support the hypothesis that exposure to coumaphos may have some beneficial effect on ability of bees to resist pathogens. They found that honey bee genes express enzymes that can detoxify harmful compounds in honey and propolis.

In addition, this year the ARS Pollinating Insect Research Laboratory in Logan, Utah, analyzed existing samples of managed solitary bees used for pollinating crops. These bees were evaluated to determine whether they were infected by any of the viruses associated with CCD in honey bees. Some of these honey bee viruses were found in both alfalfa leaf-cutting bees and alkali bees, with infection rates that ranged from 3 to 34 percent of the population.

Pennsylvania State University extension specialists are working with the USDA Agricultural Marketing Service's National Science Laboratory in North Carolina to analyze wax, pollen, and brood samples for the presence of pesticides for beekeepers who suspect colony losses associated with exposure to pesticides. The Foundation for the Preservation of Honey Bees and Project Apis m. (PAm) are contributing funds to cover half of the cost of the pesticide analysis. Information from samples is being stored in a centralized database that beekeepers and researchers can access. At this time, the database is not accessible to the general public.

EPA is collating beekill incident information provided through a variety of sources to document whether particular pesticides are associated with the losses. While pesticide manufacturers (registrants) are required to report incident data directly to the EPA, the Agency also relies on reports from State and local governments and the public to populate its incident database. EPA has increased the number of options available to the public for reporting beekill incidents directly to the Agency.

Topic III: Research to Identify Factors Affecting Honey Bee Health, Including Attempts to Recreate CCD Symptomology

Numerous research efforts jointly supported by ARS and NIFA continue to investigate factors that may play a role in causing CCD, either alone and/or in combination. Factors include diseases (parasites and pathogens), pesticides, poor nutrition, beekeeping practices, and to a lesser extent, other pests such as the small hive beetle.

Possible Mortality Factors and Their Interactions

Researchers at Pennsylvania State University conducted laboratory bioassays with specific levels of pesticides to measure a variety of effects on bees. They found synergistic interactions on bee larvae between a crop fungicide and a miticide used by beekeepers for *Varroa* control. This means that the bees suffered more serious effects when exposed to more than one pesticide at the same time and that bee management practices may be affecting colony health. Further studies using pesticide dosages equivalent to those actually used in the field (environmentally relevant concentrations) need to be conducted to show if the laboratory study reflects what actually occurs in the environment. The researchers also found that larval and adult honey bees had different sensitivities to pesticides. A fungicide that was very toxic to bee larvae did not kill adult honey bees. These studies utilized exposure methods that represent the worst case scenario and do not necessarily reflect how larval bees may actually be exposed through consumption of brood food. Additional studies demonstrated that common inert ingredients used in pesticides had a high toxicity in honey bee larvae, although the extent to which these inerts would be present under actual field conditions is yet to be determined. These preliminary studies suggest that fungicides, inert ingredients, and pesticide interactions may have crucial impacts on development and survival of honey bees, depending on the extent to which the inert may be present under the conditions of actual use.

Previously, scientists at the University of Nebraska–Lincoln documented the harmful synergistic effects of certain miticides and fungicide combinations on honey bee health in laboratory studies (Johnson, *et al.* 2010). Although further testing needs to occur at environmentally relevant concentrations, these scientists recommend that beekeepers be cautious about applying these pesticides in combination.

Beekeepers should particularly avoid applying miticides when honey bees are placed in orchards or other crop settings where exposure to fungicides is likely. Studies are underway to assess the effects of exposure to simultaneously applied miticides and fungicides at environmentally relevant concentrations on brood survival, weight gain, and queen performance.

As previously reported, Israeli acute paralysis virus (IAPV) was found to be associated with CCD and could be an important CCD indicator. Following up on this finding, ARS scientists studied the role of *Varroa* mite in IAPV transmission among honey bees. Because mites feed and move between adult bees and brood, they have the potential to act as a vector to transmit pathogens from infected bees to healthy bees, and they have been reported to be associated with viral disease outbreaks in the field.

Research results demonstrated that *Varroa* mite is a biological vector of IAPV by supporting replication of the virus and transmitting IAPV among honey bees. Therefore, controlling *Varroa* mite might be an important way to reduce the spread of IAPV in bee hives and could be a strategy for reducing bee decline.

Bumble Bee Decline

Some bumble bee species also have experienced drastic population declines, even to the point of extinction. This brings into question whether CCD is solely a honey bee issue or part of a more general problem among bees of many kinds. This past year, a North American Bumble Bee Conservation Planning meeting was organized for the first time (November 2010) to bring together researchers, government agency representatives, conservation organizations, beekeepers, and growers to discuss the recent declines in bumble bees and identify needs related to bumble bee conservation.

The meeting was facilitated by a representative from the International Union for the Conservation of Nature (IUCN) and resulted in the formation of a bumble bee specialist group to advise IUCN on bumble bee issues. This group is preparing a report of its findings.

Immune System Effects

Another study conducted by ARS scientists resulted in a somewhat counterintuitive conclusion, a finding that temperature-stressed bees were more resistant to infection by pathogens than nonstressed bees. Using alfalfa leafcutting bees (*Megachile rotundata*) as a model, experiments determined that bees exhibit a heightened biological response to temperature stress that has many similarities to the immune response.

Thus, if bees were stressed by higher temperatures before a pathogen started to invade, infection levels were lower. The study hypothesized that the pathogen invaded nonstressed bees more readily, as these bees did not have time to activate their immune response before the pathogen was able to become established and disable some of the honey bees' immune functions.

Topic IV: Mitigation and Management Preventive Measures

Trilateral Discussions to Prevent Invasive Pathogens

To meet the need for a stronger regulatory framework for controlling the accidental introduction of exotic bee pathogens, APHIS and ARS initiated trilateral discussions between Mexico, Canada, and the United States through the North American Plant Protection Organization (NAPPO), with the goal of developing coordinated guidelines for the regulation and importation of pollen. The importation of pollen as a source of food for bees is currently not allowed in the United States and Canada, but it is allowed in Mexico. However, the importation of pollen for human consumption (regulated in the United States by the Food and Drug Administration) is allowed in all three countries, which could provide a pathway for the introduction of new pathogens. The NAPPO technical advisory group is also trying to determine whether mitigation measures such as radiation or ozone are feasible. APHIS is working to amend its regulations to allow pollen importation for bee feed under permit with the appropriate safeguards.

Best Management Practices (BMP) Guide

To assist beekeepers and growers relying on pollinators, a *Best Management Practices Guide for Beekeepers Pollinating California's Agricultural Crops* (www.beeccdcap.uga.edu /documents/bmpcalagr.html) was developed by members of the NIFA-CAP team. The Guide was published in the *American Bee Journal* and in the trade magazine *Bee Culture*, two of the most widely read beekeeping publications in North America. The article offers guidance to beekeepers and almond growers in seven areas: nutrition, *Varroa* mite control, *Nosema* control, management of hive equipment, colony management, business management, and guidance to almond growers renting bees.

Genomics Tools to Manage *Varroa* Mites

A genome draft sequence was published for the *Varroa* mite, revealing potential weak points in mite biology (defensive proteins and proteins used in chemical mitigation) and candidates for novel controls such as RNA interference (RNAi). In fact, the publication of mite candidate genes in genomic databases led to worldwide control studies for this parasite using RNAi technology. Microbes also identified in this study have been screened across bees and mites as possible controls for *Varroa*.

Varroa-Resistant Bees for Commercial Use

ARS scientists tested Russian honey bees and bees containing a *Varroa*-resistant trait (*Varroa*-sensitive hygiene [VSH]) for productivity in two largescale field tests using two different migratory beekeeping routes. Two years of testing during commercial pollination of almonds, apples, blueberries, and cranberries, with overwintering in Louisiana, showed the *Varroa*-resistant stocks to be as large and productive as control stocks at each pollination site. A 1-year test involving almond pollination and Midwest honey production, with overwintering in California's Central Valley, again showed Russian and VSH bees to have adequate colony size, survivability, and honey production. The results suggest these *Varroa*-resistant bees are well suited for commercial pollination services and honey production.

EPA's Expediting of Reviews for Section 18

EPA is working with State liaison agencies to provide beekeepers with appropriate tools to control bee colony pests such as *Varroa* mites.

The Agency is working closely with USDA's ARS and OPMP to expedite reviews of Emergency Exemption requests (under section 18 of the Federal Insecticide, Fungicide and Rodenticide Act; FIFRA), Special Local Needs Registrations (under section 24c of FIFRA), and full registrations (under section 3 of FIFRA) of miticides.

Blue Orchard Bees for Almond Pollination

This past year, three workshops were conducted by collaborating scientists with the California Farm Advisors and ARS to train almond producers and blue orchard bee (*Osmia lignaria*) producers on methods for raising this bee and for using it effectively as an almond pollinator.

Research also continues to seek the best way to integrate bee hives for the most cost-effective and complete pollination.

Field trials were established this year in a joint collaboration between ARS and almond producers in California.

Bumble Bee Rearing and Investigations of Bumble Bee-Honey Bee Disease Transmission

Research continued toward developing mass rearing methods for bumble bee colonies using native western bumble bees (*Bombus occidentalis*). Eastern U.S. species are thought to carry diseases that may affect wild bumble bee populations in the Western United States, so commercially available species are needed for Western farmers.

Pollinator studies have also focused on exploring the use of bumble bees as pollinators and seeking to determine the relationship between bumble bee and honey bee pests and diseases. Studies have revealed that bumble bees are affected by IAPV and other viruses and parasites that affect the honey bee. These findings may have potential implications for the role of these factors in bee decline and may assist researchers to develop an understanding of, and ultimately controlling, bee parasites.

Enhancing Pollinator Forage

The Natural Resources Conservation Service has continued to encourage private landowners to establish or enhance pollinator forage by providing appropriate flowering plants from early spring through late fall.

The Appendix provides more results and findings from the past year of CCD research, listed within the framework of the CCD Action Plan.

APPENDIX: ACCOMPLISHMENTS BY ACTION PLAN COMPONENT

Topic I: Survey and Sample Data Collection

Pesticides have been identified as one of the factors that potentially contribute to CCD and to diminished pollinator health. The EPA's Office of Pesticide Programs (OPP), which is tasked with regulating pesticides, has been taking several actions to enhance the tools used to regulate pesticide products.

Global Workshop on Pesticide Risk Assessment for Pollinators, January 2011.

Representatives of the U.S. Department of Agriculture (USDA), including ARS and APHIS, and EPA's OPP served as panelists in a Society of Environmental Toxicology and Chemistry (SETAC) global Pellston Workshop on Pesticide Risk Assessment for Pollinators. Workshop participants totaling 48 from 5 continents were tasked with advancing the current state of the science of pesticide risk assessment by more thoroughly vetting quantitative and qualitative measures of exposure and effects on individual bees and at the colony level. The SETAC Workshop aimed to synthesize global understanding of exposure and effects measurements and work toward a common/harmonized global process for evaluating and quantitatively characterizing risk to insect pollinators from exposure to pesticides. The SETAC Workshop focused on four major topics to—

1. Design/identify testing protocols to provide an understanding of potential exposure to bees (including daily ingestion rates) from pesticide residues in pollen, and nectar, as well as from exposure through other routes of direct and indirect exposure;
2. Design/identify testing protocols to measure effects of pesticides to developing brood and adult honey bees at both the individual and colony level;
3. Propose a tiered approach for characterizing the potential risk of pesticides to pollinators; and
4. Explore the applicability of testing protocols, used for honey bees (*Apis* bees), to measure effects of pesticides and pesticide risk to native (non-*Apis)* insect pollinators.

SETAC will publish an overview (synopsis) of the Workshop online in summer 2011, and the full proceedings of the workshop will be published by SETAC in early 2012.

EPA has committed to developing a quantitative process for evaluating the potential risks of pesticides to insect pollinators for use in regulating pesticide products. EPA will examine the conclusions of the SETAC workshop and will utilize its Scientific Advisory Panel (SAP) to publicly vet a proposed process in summer 2012.

Coordination with International Partners. Efforts, such as those by EPA, have been underway to coordinate elements of science and policy with international partners such as the Organization for Economic Co-operation and Development (OECD) to better understand the potential role of pesticide use in bee declines.

The OECD Working Group on Pesticides has established the Pesticide Effects on Insect Pollinator (PEIP) Expert Group that has developed a series of activities[2] (http://www.oecd.org/dataoecd/19/27/45275778.pdf) on pollinator testing, research, mitigation, and information management related to insect pollinator declines.

The PEIP is currently working on four activities: (i) establishing means for efficient communication of pollinator incidents among regulatory authorities; (ii) developing pollinator testing requirements; (iii) sharing information aimed to mitigate potential risk of pesticides to pollinators; and (iv) developing a pollinator research clearinghouse.

As a first step, questionnaires related to these topics were distributed to OECD member countries.

Responses to these questionnaires will be used to develop tools to facilitate timely sharing and understanding of available information.

PROJECT CONTACTS
Tom Steeger (Steeger.Thomas@epa.gov)
Tom Moriarty (Moriarty.Thomas@epamail.epa.gov)

Goal 1: Determine the Extent of CCD in the United States

1. Definition of CCD: Refine CCD symptomology to determine what CCD is and what it is not. and
2. Develop and conduct an expanded, systematic, nationwide, epidemiological survey, based on existing models.

Accomplishments

U.S. Bee Loss Survey

ARS, in collaboration with the Apiary Inspectors of America, conducted a nationwide bee loss survey for the fourth consecutive year. Participation this year was at an all-time high, in part due to survey availability online. Results indicated a trend similar to previous years with this past fall-winter losses at 34.4 percent, consistent with the 33 percent average loss rate of past surveys. CCD-like symptoms continued to be identified by beekeepers as being observed in some colony losses, with 80 percent of commercial beekeepers reporting these symptoms. U.S. beekeepers continue to have elevated colony losses that include CCD in the fall and winter, which does not include the loss of colonies in the spring and summer; over time, these losses are a major economic drain on beekeepers.

PROJECT CONTACTS
Dennis vanEngelsdorp (dennis.vanengelsdorp@gmail.com)
Jeff Pettis (jeff.pettis@ars.usda.gov)

Improved Field Collection and Shipping Methods for Bee Diagnoses

A novel shipping method for live worker bees was developed and proven in field and laboratory analyses. This method improves on prior strategies in cost to shippers (often beekeepers) and in the ability to deliver material suitable for RNA analysis. It is now the method of choice for forensic analyses, including for the USDA-APHIS and USDA-ARS National Bee Survey.

PROJECT CONTACTS
Jeff Pettis (jeff.pettis@ars.usda.gov)
Dennis vanEngelsdorp (dennis.vanengelsdorp@gmail.com
Judy Chen (judy.chen@ars.usda.gov)
Jay Evans (jay.evans@ars.usda.gov)

Tool Development for Tracking and Understanding CCD

BRL scientists improved current methods to (1) collect field samples of honey bee populations and ship them for genetic analyses, (2) stabilize and extract RNA, (3) conduct high-throughput genetic screens for viruses and other pests, (4) collect embryos from established colonies, and (5) carry out controlled experiments on adult bees using sterile cups. These methods are being used in national surveys in the United States in order to establish cell lines and other genetic techniques, and in attempts to combine possible causes of CCD (pathogens and pesticides) in an attempt to determine interactive effects.

PROJECT CONTACTS
Judy Chen (judy.chen@ars.usda.gov)
Jay Evans (jay.evans@ars.usda.gov)

Goal 2: Determine Current Status of Honey Bee Colony Production and Health

1. Develop a long-term annual APHIS survey on the overall health status of U.S. honey bees.

Accomplishment

USDA-APHIS Bee Pest Survey

APHIS, in collaboration with ARS and 13 States, completed a limited pest and disease survey to look for exotic pests. The survey showed that no exotic *Tropilaelaps* mites, Slow Paralysis Virus, or *Apis ceranae* bees were detected in the 349 samples representing 2,700 colonies across the United States. *Nosema ceranae* was the only species of *Nosema* detected, and no tracheal mites were detected, although they have been common in the United States in past years. The survey results are available online at the APHIS Bee Health site. The national survey expanded to include 34 States this year.

PROJECT CONTACTS
Robyn Rose (robyn.rose@aphis.usda.gov)
Jeff Pettis (jeff.pettis@ars.usda.gov)

Topic II: Analysis of Existing Samples

Goal 1: Identify and Characterize Pathogens Associated with CCD

1. Analyze samples using—

- High-throughput sequencing for pathogen detection in individual colonies.
- Microarray analysis and quantitative gene expression studies to determine stressor or pathogen effects on bee gene expression.
- Integrated Virus Detection System (IVDS) for identifying pathogens by particle size.

and

2. Isolate, purify, and quantify microbes associated with CCD.

Accomplishments

New Viruses Discovered in Large Migratory Bee Operation

Scientists at the University of Californian–San Francisco conducted a study of a large-scale migratory beekeeping operation that transported bees from South Dakota to Mississippi and California in 2010. Using comprehensive molecular detection methods, including a custom microarray, quantitative polymerase chain reaction (qPCR), and ultra-deep sequencing, the researchers uncovered the seasonal incidence and abundance of several viruses, the microsporidian *Nosema* sp., *Crithidia mellificae*, and several bacteria that are well-known bee pathogens. In addition, they discovered four novel RNA viruses, two of which were the most abundantly observed components of honey bee pathogens. One virus, the newly named Lake Sinai virus strain 2 (LSV2), predominated.

Funding: Project Apis m., Howard Hughes Medical Institute, Genentech Graduate Student Fellowship, A.P. Giannini Foundation Medical Research Fellowship, and UC-Davis's Häagen-Dazs Postdoctoral Fellowship in Honey Bee Biology.

PROJECT CONTACT
Joe Derisi (joe@derisilab.ucsf.edu)

Presence of Chronic Bee Paralysis Virus in the United States

Chronic bee paralysis virus (CBPV) is a disease agent that causes characteristic "paralysis" symptoms in adult honey bees and could lead to host and colony mortality. Using developed molecular diagnostic tools, ARS scientists presented the first evidence of CBPV infection in honey bees in the United States. The study yielded important information on the incidence of CBPV infection in the U.S. population of adult honey bees and demonstrated the phylogenetic relationships of U.S. strains of CBPV with isolates of CBPV from the different geographical regions of the world, thereby increasing our understanding of virus diversity in the United States.

PROJECT CONTACTS
Judy Chen (judy.chen@ars.usda.gov)
Jay Evans (jay.evans@ars.usda.gov)

Bumble Bees and Israeli Acute Paralysis Virus (IAPV) and Deformed Wing Virus (DWV) Studied

Researchers found that IAPV infects bumble bees and reduces colony life span. Bumble bees dying from IAPV infections returned to the colony to die inside the hive, unlike honey bees, which died outside the hive. Bumble bees, however, also exhibited the same paralytic-like seizures as observed in honey bees.

Funding: USDA-NIFA.
PROJECT CONTACTS
Anne Averill (aaverill@ent.umass.edu)
Lee Solter (lsolter@illinois.edu)
Frank Drummond (frank.drummond@umit.maine.edu)

A New Bee Fungus Found in Association with Solitary Bees

A new fungus that is closely related to the fungal pathogen that causes chalkbrood (*Ascosphaera apis*) was found in the pollen stores and nesting materials of alfalfa leafcutting bees, *M. rotundata*, in Canada and the western United States.

The new fungus was named *A. subglobosa*. This new species, closely related to *A. atra* and *A. duoformis*, is distinct from other *Ascosphaera* species by its evanescent spore balls, globose to subglobose spores, and unique nuclear ribosomal DNA ITS sequence.

Discoveries such as this non-pathogenic *Ascosphaera* are assisting in elucidating the evolution of microbes to the pathogenic state, and could help us better understand the sudden arrival of new bee pathogens.

Funding: USDA-ARS and University of Copenhagen.

PROJECT CONTACT
Rosalind James (rosalind.james@ars.usda.gov)

Characterizing Chalkbrood Strains in Honey Bees

The pathogenic fungus *A. apis,* which causes chalkbrood disease, is ubiquitous in honey bee populations. The draft genome assembly of this pathogen was used to search for DNA markers adequate to identify different strains and populations of this pathogen. DNA primers were designed for five different markers and tested against a panel of closely related fungal species. These markers were compared for 12 distinct isolates of the pathogen. Use of three of the most variable markers had the same detection power as use of all five markers.

Funding: USDA-ARS and University of Copenhagen.
PROJECT CONTACT
Rosalind James (rosalind.james@ars.usda.gov)

Non-Apis Bee Pathogens Identified and Characterized

University scientists have been studying several parasites associated with bumble bees (i.e., *Crithidia bombi;* a protozoan parasite of bumble bees), *Nosema bombi*, and the tracheal mite in order to determine their transmission patterns, occurrence and distribution, and origin. Although the protozoan was found in commercial bumble bees, it does not seem to be transmitted from wild bumble bees to managed honey bees in the area of Massachusetts investigated. Results from these studies will assist in understanding and ultimately controlling bee parasites.

Funding: USDA-NIFA.
PROJECT CONTACTS
Anne Averill (aaverill@ent.umass.edu)
Lee Solter (lsolter@illinois.edu)
Frank Drummond (frank.drummond@umit.maine.edu)

Goal 2: Identify and Characterize Pests Associated with CCD

1. Use standard sampling methods to analyze samples for tracheal and *Varroa* mites and *Nosema* species.

Accomplishment

Interactions between Pests, Pathogens, and Pesticides Being Investigated

With funding from the NIFA Coordinated Agricultural Project, scientists are examining the potential interactive effects of pests, pathogens, and pesticides in seven States across the United States. After the first 2-year trial, preliminary results indicate that surrounding landscape, pesticide exposure, parasites, and diseases were significant factors in explaining overall colony losses, and not losses due to CCD alone. As the percentage of intensive agriculture in the landscape surrounding an apiary site increased, it was found that colony losses in the spring also increased. This might be due to decreased flower resources, but it also could be related to increased pesticide exposure. Additional research is needed to resolve this uncertainty. It was observed that as pesticide-contaminated pollen brought back to the hive increased, the queen replacement due to supercedure also increased, suggesting a potential mechanism for overall colony losses. However, the laboratory also found that colony losses during the spring and summer buildup season were due to the parasitic *Varroa* mite, *Nosema* disease, and IAPV. Overwintering colony losses were found to be a function of *Varroa* levels and IAPV in colonies. Although further studies are needed to make firm conclusions, these results suggest that CCD might be a combination of *Varroa*, *Nosema*, and viral disease. In addition, these results corroborate the findings of previously published genomic and proteomic analyses suggesting *Varroa* and viral associations for CCD. The relationship between pesticide exposure to bees and queen supercedure suggests that pesticides also could play a role in weakening colonies that ultimately might lead to losses.

Funding: USDA-NIFA.

PROJECT CONTACTS
Frank Drummond (frank.drummond@umit.maine.edu)
Nancy Ostiguy (nxo3@psu.edu)
Katherine Aronstein (kate.aronstein@ars.usda.gov)
Steve Sheppard (shepp@wsu.edu)
Kirk Visscher (visscher@ucr.edu)
Marla Spivak (spiva001@umn.edu)
Jamie Ellis (jdellis@ufl.edu)
John Adamczyk (john.adamczyk@ars.usda.gov)
Steve Cook (steve.cook@ars.usda.gov)

Goal 3: Identify pesticides or environmental contaminants associated with CCD

1. Examine wax, pollen, honey, and adult bee samples for pesticides and environmental contaminants.

Accomplishments

Collecting and Assessing Pesticide-Related Bee Kill Incidents

In its role in regulating pesticides, EPA requires data to document the potential effects of pesticides on honey bees at the level of the individual organism and at the colony level, depending on the outcome of lower tier tests and other scientifically relevant information. EPA also collects and assesses data on ecological (bee kill) incidents associated with the use of pesticides. Incidents are typically reported to State Lead Agencies (SLAs; consisting primarily of State departments of agriculture) within each of the 50 States, where formal investigations are conducted to determine whether specific pesticides may have been associated with the incidents. Incident reports may also be reported to pesticide registrants who may also conduct investigations. Only the pesticide registrants are required by law to report incidents to EPA; States that investigate incident reports may or may not relay the results of those investigations to EPA. Incident reports received through this process are recorded in a database (i.e., the Ecological Incident Information System [EIIS]) maintained by OPP.

In addition to this formal process for reporting and investigating incidents, incidents may also be reported by the public through the National Pesticide Information Center (NPIC) Web site (http://npic.orst.edu /reportprob. html#env), maintained by Oregon State University under an assistance agreement with EPA. Bee kill incident reports may also be submitted directly to EPA via e-mail using the beekill@epa.gov address that is available on EPA's pollinator Web site (http://www.epa.gov/opp00001/ecosystem /pollinator/science.html) or by contacting the Agency directly by phone. Incident reports that are not investigated, however, may not contain sufficient information to support risk assessments that rely on these data as a line of evidence for whether a chemical can impact non-target species such as honey bees. Detailed information (such as pesticide use information, residue data, number of bees/colonies lost, time over which the incident took place, and colony health/management) enables risk assessors to better evaluate whether a particular pesticide can be reasonably linked to a bee kill incident.

PROJECT CONTACTS
Tom Steeger (Steeger.Thomas@epa.gov)
Tom Moriarty (Moriarty.Thomas@epamail.epa.gov)

Pesticide Variability Analyzed in Stationary Hive Study

Scientists supported by NIFA's Coordinated Agricultural Project analyzed pesticide residues in stationary bee hives and identified a variety of different chemicals and their metabolites in hives. During the past 2 years (2009–2010), these researchers have been monitoring pesticides in pollen collected by foraging honey bees at the apiaries set up for the stationary hive survey in both rural and urban environments. The amount of pesticides observed varies with time and location. On average, each sample has had residues of four pesticides. A total of 45 different pesticide compounds or metabolites have been observed during the past 2 years including insecticides, fungicides, and herbicides.

Funding: USDA-NIFA.
PROJECT CONTACTS
Brian Eitzer (Brian.Eitzer@ct.gov)
Frank Drummond (frank.drummond@umit.maine.edu)
Nancy Ostiguy (nxo3@psu.edu)
Katherine Aronstein (kate.aronstein@ars.usda.gov)
Steve Cook (steve.cook@ars.usda.gov)
Steve Sheppard (shepp@wsu.edu)
Kirk Visscher (visscher@ucr.edu)

Pesticide Exposure to Bees Documented in 1100 Samples Across the United States

Scientists from Pennsylvania State University and ARS continue to document both unintentional and intentional exposure of honey bees to pesticide residues in hive products, especially beebread and beeswax. Samples were collected from migratory beekeepers in 23 States and one province in Canada. These samples included bees that were collected during beekill incidents. Honey bees and possibly other pollinators were exposed to a large number (and often high levels) of pesticides, as suggested by an average of 6.2 pesticides per pollen sample and up to 31 pesticides in a single sample. More than 1,100 samples of bee products were analyzed to date, with 130 different pesticides and metabolites detected from North American apiary samples alone. The frequency and quantity of pesticides detected varied considerably, and most of the samples were close to the level of quantification for their respective pesticides. Particularly noteworthy is the fungicide chlorothalonil, which was found in the majority of samples analyzed. Almost all comb and foundation wax was contaminated with miticides used to control parasitic mites of honey bees and other pesticides, averaging 8 detections with a high of 39 per sample; however, again the quantity of pesticides detected varied considerably. Although no single pesticide was shown to be responsible, the additive and synergistic effects of multiple pesticide exposures are believed to be contributing factors in declining honey bee health, but further studies are needed to better support this hypothesis. Potential pesticide interactions with other stressors including *Varroa*, IAPV, and *Nosema*, and its role in CCD require further study. In the future, researchers will focus on impacts of multiple pesticide residues in bee food via synergistic interactions that can act at sublethal levels on key pollinator behaviors/physiology including memory and learning, and on immune function. Results of these studies were published (Mullin *et al.*, 2010 PLoS ONE 5: 1 19e9754).

Funding: USDA-NIFA, Pennsylvania State University, and National Honey Board.

PROJECT CONTACTS
Chris Mullin (CAMullin@psu.edu)
James L. Frazier (jff2@psu.edu)
Maryann Frazier (mxt15@psu.edu)
Dennis vanEngelsdorp (Dennis.Vanengelsdorp@gmail.com)
Jeff Pettis (jeff.pettis@ars.usda.gov)

Cost-sharing Program for Pesticide Analysis of Honey Bee Colony Matrices (Honey, Wax, Pollen, Bees, Brood, etc.)

Based on recent evidence of frequent pesticide detections in wax, pollen, and brood, beekeepers have expressed an interest in having samples from their own colonies/apiaries tested for pesticides. A fund has been established to assist beekeepers by paying half of the cost of the pesticide analysis. Analysis is being performed by the USDA AMS National Science Laboratory in Gastonia, North Carolina, and the information generated from individual samples becomes part of a large, centralized database maintained at Pennsylvania State University. At this time, the database is not accessible to the general public.

Funding: Foundation for the Preservation of Honey Bees and Project Apis m. (PAm).

PROJECT CONTACT
Maryann Frazier, Pennsylvania State University (mxt15@psu.edu)

Reduction of Honey Bees Pollinating Nine Crops

In 2009 and 2010, researchers assessed changes in nine colonies on each of nine field crops by counting foragers exiting colonies during the pollination period for each crop. Researchers also conducted pesticide analyses on samples of dead and dying bees around hive entrances, returning foragers, crop flowers, trapped pollen, and corn flowers located near a cotton crop. Colony levels were significantly reduced in hives pollinating cotton, corn, and alfalfa, while hives placed in apples, pumpkins, almonds, melons, blueberries, and wild flowers (for honey production) increased or remained fairly consistent. A total of 52 pesticide residues was detected in samples collected across the 9 crops. Pesticide residues were not detected on the target crops (those crops in which the bees were rented for pollination) but were found in trapped pollen in hives or on dead and dying bees in the vicinity of the hives. It is difficult to say with certainty where the residues came from, but bees were possibly visiting other crops or plants treated with the pesticides found in trapped pollen or on the dead and dying bees. Since honey bees require diverse sources of pollen, the establishment of flowering, pesticide-free plants near bee-pollinated crops could help mitigate exposure to pesticides.

Funding: National Honey Board and USDA-NIFA.

PROJECT CONTACTS
Maryann Frazier (mxt15@psu.edu)
Chris Mullin (CAMullin@psu.edu)
James L. Frazier (jff2@psu.edu)

Movement of Imidacloprid in Trees Explored

APHIS has used soil and trunk injection of imidacloprid to treat hardwood trees in controlling the invasive Asian long- horned beetle. There is the potential for movement of the pesticide into pollen and nectar of treated trees. APHIS is collaborating with ARS to monitor the movement of imidacloprid into pollen and nectar and to evaluate the potential effects on honey bees and other non-target species over a 3-year period.

Low levels of imidacloprid were found in flowers and pollen during the initial 3 years of this study. There were greater amounts of imidacloprid found in pollen from soil-injected than trunk-injected trees, and male flowers had higher amounts than female flowers. However, little of the imadicloprid found in flowers was from pollen, and no metabolites of imidacloprid were found in pollen. Although there were instances where sublethal levels of imidacloprid to honey bees were found in pollen from treated trees, this pollen is probably diluted and mixed with other pollen in the hive, thus reducing potential exposure to sublethal doses.

This study has been extended for 3 years to further monitor for and quantify imidacloprid residue levels in flowers, pollen, nectar, and leaves; determine the impact of residues on honey bee health; and compare results of soil- and trunk-injected trees.

Funding: USDA-APHIS and USDA-ARS.

PROJECT CONTACTS
Jeff Pettis (jeff.pettis@ars.usda.gov)
Robyn Rose (Robyn.I.Rose@aphis.usda.gov)

Ongoing Research

Stationary Apiary Health Parameters Comparisons

In the NIFA Coordinated Agricultural Project (CAP), scientists assessed the presence and quantity of tracheal and *Varroa* mites, viruses, *Nosema* in bees and pesticides in wax, as well as quantity of pollen, queen health parameters, and brood and adult bee populations, evaluating possible linkages between pest/pathogen/pesticide presence and colony survivorship. Data show high variability among virus presence, and evaluations continue to identify possible linkages.

There was also a high variability in the pesticide load in the pollen brought back to the hives.

Funding: USDA-NIFA.

PROJECT CONTACTS
Nancy Ostiguy (nxo3@psu.edu)
Brian Eitzer (Brian.Eitzer@ct.gov)

Goal 4: Determine bee physiological response to toxins, pesticides, or pathogens and develop analytical tools to assess bee health

1. Compare genes expressed in response to specific pathogens or pesticides with those expressed in bees from CCD colonies,
   and
2. Develop the use of molecular markers to determine the physiological status of bees and as indicators of bee health.

Accomplishments

Residue Analysis Yields Clues to Potential Fate of Pesticides

EPA OPP's Analytical Chemistry Laboratory in Fort Meade, Maryland, provides residue analysis of neonicotinoid pesticides in honey bee colonies in support of research being conducted by USDA in conjunction with the University of Maryland.

These studies are informing EPA's understanding of the potential exposure of honey bees to pesticide (neonicotinoid) residues in pollen and nectar. In addition, these studies have provided information on the potential effects of imidacloprid on honey bee colonies following prolonged exposure to the compound in diets (pesticide-spiked bee bread).

In support of pesticide residue analyses, EPA's Analytical Chemistry Laboratory has refined analytical techniques used for measuring residues of imidacloprid and its degradates and has increased the detection limits for these residues. These new methods have been published in the open literature.

Funding: EPA.

PROJECT CONTACTS
Tom Steeger (Steeger.Thomas@epa.gov)
Tom Moriarty (Moriarty.Thomas@epamail.epa.gov)

Methods Developed to Detect Virus Replication in Honey Bees

The assessment of virus effects on bee colony health requires not only the detection of viruses in bees and bee colonies but also the indication of virus replication in the host.

Scientists at the University of Massachusetts have developed a simple, real-time polymerase chain reaction (PCR) based assay to detect the most active stage of virus replication—production of a material known as "negative strand, replicative intermediate"—and determined that 50–80 percent of the virus-positive bees had these intermediates. These results indicate that actual virus infection rates were much lower than previously understood.

Funding: USDA-NIFA.

PROJECT CONTACT
John P. Burand (jburand@microbio.umass.edu)

Gene-Expression Analyses Using Candidate Immune- and Stress-Response Genes

Gene expression analyses with parallel bee gene sets and pathogen genes have revealed changes in the set of expressed bee genes that could relate to the cause(s) of collapsing colonies, including a massive shift in the abundances of gene transcripts related to protein translation.

These experiments were complemented by high-throughput sequencing of millions of honey bee RNAs for bees challenged by the American foulbrood bacterium (*Paenibacillus larvae*) and bees from CCD and control colonies.

Funding: USDA-ARS.

PROJECT CONTACTS
Jay Evans (jay.evans@ars.usda.gov)
Scott Cornmann (scott.cornmann@ars.usda.gov)

Genes Associated with Nosema Infection Determined by Microarrays

In 2010, we verified the status of bees infected with *Nosema* using DNA and microarray approaches.

Microarray analyses revealed that *Nosema* infection alters host metabolic pathways regulating nutrition and behavioral maturation as expected, but *Nosema* infection surprisingly does not appear to significantly alter immune gene expression in midgut and fat body tissues up to 7 days post-infection.

We will continue to examine impacts of infection by characterizing gene expression in immune-related tissues up to 2 weeks post-infection in bees infected with *Nosema*. These studies will identify host response to *Nosema* infection and may lead to downstream applications in commercial management.

Funding: USDA-ARS and USDA-NIFA.

PROJECT CONTACTS
Katherine Aronstein (kate.aronstein@ars.usda.gov)
Christina Grozinger (cmgrozinger@psu.edu)

Diagnostic Tools Developed for High-Throughput Detection and Monitoring of Honey Bee Diseases

ARS scientists developed a new diagnostic tool for detection of *Nosema* infection in field honey bee samples. This highly specific, sensitive test, based on an antigen capture assay that detects *Nosema* spore wall protein, is capable of detecting *Nosema* antibodies. Researchers are pursuing leads to transfer this technology to private industry. These findings have been accepted for publication in the *Journal of Apicultural Research.*

Following the development of a reliable field diagnostic tool, ARS scientists are now using the same antibody-based approach to develop a high-throughput laboratory test (enzyme-linked immunosorbent assay; ELISA) for *Nosema* detection in bees. In response to increased incidence of fungal diseases in bee colonies, ARS scientists developed a simple DNA-based method for detection of chalkbrood fungal disease in bee brood. This test will allow detection of the fungus prior to clinical signs of the disease. Although chalkbrood disease is usually not a primary concern for the beekeepers, it has recently become more prevalent. Research indicates that high levels of stress as a result of intense management of the environment may affect bee immune responses, making them more vulnerable to what were once considered benign diseases.

Funding: USDA-ARS and USDA-NIFA.

PROJECT CONTACTS

Katherine Aronstein (kate.aronstein@ars.usda.gov)
John Adamczyk (john.adamczyk@ars.usda.gov)

Topic III: Research to Identify Factors Affecting Honey Bee Health, Including Attempts to Recreate CCD Symptomology

Goal 1: Confirm or Eliminate Potential Environmental Stressors as Contributing Causes of CCD

1. Test effects (lethal and sublethal) of neonicotinoids and other pesticides used for crop protection.

Various research results indicate the presence of a range of pesticides and their degradates in honey bee colonies; however, no research or survey information has linked the incidence of CCD to pesticides. This is consistent with the multi-factorial hypothesis being pursued by the CCD Steering Committee. To the extent that pesticides are demonstrated to be directly impacting honey bee survival, growth and/or reproduction, EPA is prepared to mitigate these effects to the extent possible under the law.

As discussed earlier, EPA is also working with its international partners through OECD to identify high-quality research on the potential effects of pesticides on pollinator health, particularly as it relates to CCD.

EPA is continuing its efforts to collaborate with researchers in both government and non-government institutions regarding the data that can potentially inform EPA's regulatory decision-making. EPA scientists have also served as *ad hoc* reviewers of research proposals related to the potential effects of pesticides on insect pollinators submitted to USDA for funding.

PROJECT CONTACTS
Tom Steeger (Steeger.Thomas@epa.gov)
Tom Moriarty (Moriarty.Thomas@epamail.epa.gov)

Accomplishments

Lethal and Sublethal Effects of Insecticides on Non-Apis Bees Elucidated

Using totally enclosed field cage systems, University of Massachusetts scientists assessed the impact of imidacloprid in flowering low bush blueberry on commercial bumble bees (*Bombus impatiens*). After bloom ended, the bees were released to forage on surrounding plants for the remainder of the spring and summer. Results suggest that imidacloprid reduced brood (immature bees) at the end of bloom but did not seem to affect the survival of adult workers.

Funding: USDA-NIFA.
PROJECT CONTACT
Frank Drummond (frank.drummond@umit.maine.edu)

Persistence of the Effect of the Insecticide Novaluron on Bee Reproduction

In a study on alfalfa leafcutting bees in field cages, an ARS scientist evaluated the persistence of toxicity of the insecticide novaluron. The effect of female bee exposure to novaluron-treated alfalfa was found to persist for at least 2 weeks post-spray; the effect was not adult mortality, but instead, mortality in the eggs laid by exposed adult females. Egg mortality was higher in cages with treated alfalfa compared to cages with untreated alfalfa. The effect diminished over time but remained higher than control mortality throughout the study.

Funding: USDA-ARS and USDA-NIFA.
PROJECT CONTACTS
Theresa Pitts-Singer (theresa.pitts singer@ars.usda.gov)
James Barbour (jbarbour@uidaho.edu)

Synergistic and Sublethal Impacts of Pesticides Determined on Honey Bee Larvae

The levels of pesticides detected within the hive have increasingly raised concerns about bee health and development under chronic exposure to pesticides. Scientists at Pennsylvania State University have used a modified *in vitro* larval rearing technique to assess the survival of honey bee larvae during a 6-day exposure to sublethal doses of four of the most frequently found pesticides in bee pollen. A Toxicity Probabilistic model was developed to determine relative toxicity of a single pesticide to honey bee larvae, which estimates pesticide toxicity as a function of the magnitude of toxicant exposure and a bee's sensitivity to a toxicant. Modeling results demonstrate that honey bee larvae and adults have different sensitivity to pesticides, and chlorothalonil, a fungicide which is considered non-toxic to bees on an acute exposure basis and is applied during bloom, has the highest relative toxicity to bee larvae compared to adult bees among the pesticides tested: fluvalinate, coumaphos, and chlorpyrifos. In addition, a method for assessing mixture toxicity was developed. Synergistic interactions were found between the fungicide chlorothalonil and the miticide fluvalinate at concentrations 34 ppm and 3 ppm, respectively, increasing larval mortality at day 4, approximately 7 times greater than the expected toxicity for mixtures with only concentration addition effect. However, when decreasing the concentration for fluvalinate and chlorothalonil by 10 fold, the mixture displayed antagonistic interaction. Chlorothalonil and coumaphos in mixtures at sublethal doses also displayed synergistic toxicity 5 times greater than the expected mortality if there is no interaction. The other binary mixtures, including fluvalinate and coumaphos, fluvalinate and chlorpyrifos, and coumaphos and chlorpyrifos, showed an additive effect. For three-component mixtures, only one significant reduction in larval mortality was found when adding coumaphos into the mixture of fluvalinate and chlorothalonil. Adding 34 ppm chlorothalonil into the three-component mixture of fluvalinate, coumaphos, and chlorpyrifos, the four-component mixture showed an additive effect. Moreover, considering pesticide formulations, the researchers demonstrated high toxicity of a common inert ingredient N-methylpyrrolidone (NMP) to honey bee larvae; NMP can kill all the reared larvae during the first day. This study suggests that fungicides, inert ingredients, and pesticide interactions may have crucial impacts on development and survival of honey bee larvae. However, it is uncertain as to the extent that formulation inerts remain with their active ingredients, and additional research is needed to resolve this uncertainty.

Funding: USDA-NIFA.

PROJECT CONTACTS
James L. Frazier (JFrazier@psu.edu)
Chris Mullin (oy9@psu.edu)
Maryann Frazier (mxt15@psu.edu)

2. Test the effects of current miticides used in hives on worker bee longevity and colony health.

Accomplishments

Interactive and Sublethal Effects of In-hive Miticides Evaluated

Scientists at the University of Nebraska–Lincoln have been studying the synergistic effects of various varroacides and fungicides on honey bee health, testing for lethal and sublethal effects. Research indicated that combinations of varroacides that are detoxified by the cytochrome P450 system (e.g., the miticides *tau*-fluvalinate, coumaphos, and fenproximate) tend to be significantly more toxic when applied in combination.

A similar interaction occurred when bees were exposed concurrently to a cytochrome P450-inhibiting fungicide and a varroacide that was metabolized by P450s. Normally these varroacides are well tolerated by honey bees, but pre-treatment with a P450-inhibiting fungicide greatly increases mortality in bees by a synergistic ratio of nearly 2000× in the most extreme case. While this interaction was found using fungicide doses substantially higher than would likely occur through consumption of fungicide-contaminated pollen (Mullin *et al.*, 2010, PLoS ONE 5: e9754), the potency of this interaction raises concerns about lethal and sublethal effects that may be occurring at field-relevant doses. The results suggest that honey bees should not be treated with more than one P450-detoxified varroacide concurrently or sequentially because varroacides are known to accumulate in beeswax. They also suggest that beekeepers should avoid applying P450-detoxified varroacides when honey bees are placed in orchards or other crop settings where exposure to P450-inhibiting fungicides is likely.

Additional studies are needed to resolve this uncertainty. These results may be useful for incorporation into a bee health management plan. Additional studies are needed to determine the range of bee exposure to fungicides and varroacides in the field and how these concentrations affect the bees.

Funding: USDA-NIFA.

PROJECT CONTACT
Marion Ellis (mellis3@unl.edu)

Further Studies on Sublethal Effects of In-hive Miticides

Previously, scientists at the University of Nebraska–Lincoln documented the harmful synergistic effects of certain varroacide and fungicide combinations on honey bee health in laboratory studies (Johnson, *et al.*, 2010). Their results suggest that beekeepers should avoid applying these varroacides in combination. Beekeepers should also avoid applying these varroacides when honey bees are placed in orchards or other crop settings where exposure to prochloraz is likely. Studies are underway to assess the effects of exposure to simultaneously applied field-relevant doses of miticides and fungicides on brood survival, weight gain, and queen performance.

Funding: USDA-NIFA.

PROJECT CONTACT
Marion Ellis (mellis3@unl.edu)

Certain Varroacides Could Actually Lower Bee Susceptibility to CCD

Pennsylvania State University and ARS researchers conducted a classification and regression tree (CART) analysis to better understand the relative importance of different risk variables in explaining CCD (van Engelsdorp *et al.,* 2010). Fifty-five variables, which included various pathogens, pesticides, and pests, were used to construct the models. Six of the 19 variables having the greatest discriminatory value were pesticide levels. Notably, coumaphos (a miticide commonly used by beekeepers) levels in brood had the highest discriminatory value and were highest in control (healthy) colonies. This suggests that higher levels of coumaphos could help colonies fight off CCD. This result must be interpreted with caution, however, because *Varroa* mite levels were similar in both CCD and non-CCD (control colonies). Follow-up studies are necessary to quantify *Varroa* infestations at different time intervals to give more conclusive results.

Funding: National Honey Board; USDA-ARS; North Carolina Agriculture Foundation; North Carolina Department of Agriculture and Consumer Services; USDA-NIFA; and University of Liege, Belgium.

PROJECT CONTACTS
Dennis vanEngelsdorp (Dennis.Vanengelsdorp@gmail.com)
Jim Frazier (jff2@psu.edu)

3. Test the effects of antibiotics (especially new ones such as Tylosin) on the increase in pathogens (e.g., *Nosema ceranae*) and the overall viability of bees over winter.
   See Topic IV, Goal 5.
4. Test effects of supplemental protein and carbohydrate (e.g., high fructose corn syrup [HFCS]) feedings on bee health.
   See Topic IV, Goal 1, Objective 1.
5. Test effects of availability and quality of natural food sources on bee health as affected by climatic factors (e.g., drought).
   and
6. Test effects of management practices (e.g., nutrition and migratory stresses) on bee health.

These objectives are being done as part of the work on migratory beekeeping and the Area-wide project. See Topic IV, Goals 1 and 6.

Goal 2: Confirm or Eliminate Potential Pathogens as Contributing Causes of CCD

1. Test pathogenicity of the following CCD-associated microbes against honey bees and non*Apis* bees:

- Viruses
- Fungi (chalk brood; stonebrood)
- Microsporidia (Nosema)
- Bacteria
- Trypanosomes and other microbes

Accomplishments

Differential Susceptibility of Bees to IAPV

Honey bee workers of different ages were found to be differentially susceptible to IAPV infection. Newly emerged adult bees were highly susceptible to infection by IAPV, with median mortality occurring at 46 hours post emergence. The susceptibility declines as bees age, with 5- to 6-day-old bees exhibiting 40 percent mortality at 80 hours. Bees parasitized as pupae (with normal wings) by *Varroa* are approximately twice as susceptible to IAPV infections as non-parasitized bees.

Funding: USDA-NIFA.

PROJECT CONTACT
Diana Cox-Foster (dxc12@psu.edu)

Honey Bee Viruses Found in Non-Apis Bees Essential to Alfalfa Seed Industry

A survey of alfalfa leafcutting bees (ALCB) (*Megachile rotundata*) and alkali bees (*Nomia melanderi*) from Utah and Washington State, and subsequent molecular analysis for seven different RNA viruses, revealed the presence of deformed wing virus (DWV), Israeli acute paralysis virus (IAPV), black queen cell virus (BQCV), and sac brood virus (SBV). Further, when IAPV was fed to virus-free ALCB larvae, the viral infection disrupted adult diapause, with IAPV-infected bees having a significantly greater percentage emerging as second-generation adults as compared to the control treatment. IAPV infection also significantly reduced the ALCB adult survivorship in both males and females when acquired in newly emerged adults. IAPV was confirmed to be replicating in both the adult and larval stages, and the infection was confined to the gut in larvae.

Funding: USDA-ARS, USDA-NIFA, and Pennsylvania Department of Agriculture.

PROJECT CONTACTS
Diana Cox Foster (dxc12@psu.edu)
Rosalind James (rosalind.james@ars.usda.gov)

Goal 3: Confirm or Eliminate Pests as Contributing Causes of CCD

1. Test the effects of *Varroa* mites on bee health and robustness, particularly overwintering effects and association with CCD in early spring.
   See Topic III, Goal 4.
2. Determine the importance of *Varroa* as a vector of viruses associated with CCD or as a general immunosuppressive agent on the colony itself.

Accomplishment

Varroa Mites and Virus (IAPV) Transmission

IAPV was identified to be strongly associated with CCD and is considered to be a significant marker for CCD. Because mites feed and move between adult bees and brood, they have the potential to act as a vector to transmit pathogens from infected bees to healthy bees and have been reported to be associated with viral disease outbreaks in the field. Research results demonstrated that *Varroa* mite is a biological vector of the virus by supporting replication of IAPV and transmitting IAPV among honey bees. The results also showed that the mite-virus association could possibly reduce host immunity and therefore promote elevated levels of virus replication. This study sheds light on the epidemiology of IAPV infection in honey bees and adds additional importance to the control of *Varroa* mites.

PROJECT CONTACTS
Judy Chen (judy.chen@ars.usda.gov)
Jay Evans (jay.evans@ars.usda.gov)

Also see Topic III, Goal 4.

Goal 4: Determine What Factors (or Interactions between Factors) Are Most Important in Their Contribution to CCD. This Includes Environmental Factors (E.G., Temperature, Humidity, and Chemical Exposure), Pathogens and Parasites, and Bee Genetics and Breeding

Accomplishments

Synergism Found between Pesticides on Honey Bee Homing Behavior

Researchers at Michigan State University determined that the distance of bee search flights and the length of time homing memories are retained by bees. An assay was developed to determine the effects of pesticides on the memories of foraging bees. In this study, bees were first released 0.8 km from the hive, and the bees returning to the hive were counted (homing rate). Those bees that were released the first time were recaptured and re-released 1.8 km away from the hive. Bees returned home equally well from the first release site (0.8 km). However, when the returning bees were re-released 1.8 km away from the hive, bees that were treated with a combination of fluvalinate and imidacloprid showed a significant reduction in their homing rate. Further, their homing rate was not affected when either pesticide was used alone. These results suggest that these two pesticides have a synergistic effect on honey bee learning and/or memory. These results shed light on a possible effect of CCD on bees, because foragers are having trouble finding their way home.

Funding: Michigan State University, Australia National University, and USDA-NIFA.

PROJECT CONTACT
Zachary Huang (bees@msu.edu)

Synergistic Effects of Pesticide Exposure on Pathogen Growth and Bee Health Discovered

There has been significant concern about possible interactive effects of various pesticides and pathogens on bees. When bees were exposed to the gut parasite *Nosema* and the insecticide imidacloprid, researchers found evidence of an interactive effect on honey bee health. Worker bees exhibited up to a fourfold increase in *Nosema* levels when they originated from colonies that had been fed imidacloprid, indicating a subtle sublethal interaction between pesticides and pathogens. The same fourfold increase was exhibited in bees exposed to 2.5 ppb or 20 ppb imidacloprid. Also, while the individual bees contained higher spore loads, the colonies from which they were derived did not exhibit elevated *Nosema* loads. This research has been submitted for publication, and similar findings have recently been published from a study in France in *Environmental Microbiology 2009*.

Funding: USDA-ARS and North American Pollinator Protection Campaign (NAPPC).
PROJECT CONTACTS
Jeff Pettis (jeff.pettis@ars.usda.gov)
Galen Dively (Galen@umd.edu)
Dennis vanEngelsdorp (Dennis.Vanengelsdorp@gmail.com)

Interactions among the Microsporidia Nosema apis and Nosema ceranae, the Viruses IAPV, DWV, BQCV, and Rearing Temperature

Development, virulence, infectivity (IC_{50}), spore production, temperature effects, and direct competition in the same host were evaluated for *N. ceranae* and *N. apis*. No significant difference in virulence between the two species was detected, but $_{IC50}$ studies demonstrated that a 5x higher dosage of *N. ceranae* is needed for infection. Host mortality at different temperatures was not significantly different for the two pathogens. *N. ceranae* produced slightly more spores than *N. apis* over the infection period when 200× the infective dosage was fed. When individual bees were inoculated with both *Nosema* species to produce mixed infections, *N. apis* produced more spores. When viruses (DWV, IAPV, and BQCV) were combined with microsporidia at different temperatures, mortality trended higher but was not significantly different from mortality due to the microsporidia alone. There was no difference in spore production by *N. apis* and *N. ceranae* when mixed with viruses. Mortality of virus and *Nosema* combinations was highest at low and high rearing temperatures, no matter the combination (and including controls), so rearing temperature was the mortality factor, not mixes of virus and *Nosema*.

Funding: USDA-NIFA and USDA-ARS.

PROJECT CONTACTS
Lee Solter (lsolter@illinois.edu)
Judy Chen (judy.chen@ars.usda.gov)
Jay Evans (jay.evans@ars.usda.gov)

Association of Miticides with Adult Bee Susceptibility to Nosema

A significant relationship between pesticide residue exposure (in brood comb) and adult susceptibility to infection with *Nosema ceranae* was detected in a common hive environment study in 2010. The three most frequently observed pesticides were miticides used to control *Varroa* mite in brood comb.

That is, adult honey bees that were reared in "high pesticide residue comb" (taken from existing commercial beekeeping operations) were significantly more likely to become infected with *Nosema ceranae* than those reared in low residue comb, suggesting a detrimental sublethal effect of pesticide exposure on the immune system. These findings were reported at several beekeeper local, regional, and national meetings. A manuscript was submitted to the *Journal of Invertebrate Pathology*. A follow-up large-scale field study was initiated in 2010 and will be completed in 2011.

Funding: USDA-NIFA-CAP; the State Beekeeper Associations of Washington, Oregon, Idaho, and California; and Project Apis m.

PROJECT CONTACT
Steve Sheppard (shepp@wsu.edu)

Interaction between Pesticides and Bee Viruses Found

Pennsylvania State University researchers found that consumption of sublethal levels of certain pesticides elevates viral titers in infected bees and appears to alter the immune responses of the bees.

In cage studies, nurse or house bees were fed sugar solutions containing pesticides (myclobutanil, acetamiprid, fluvalinate, or chlorothalonil) at dosages previously determined to be sublethal.

These dosages approximated lower concentrations found in incoming pollen. The titers of several picorna-like viruses (deformed wing virus, black queen cell virus, and IAPV) were found to be elevated by one or several of the pesticides.

The expression of several immune-related genes was altered by the pesticide exposure as compared to untreated bees that were also immune challenged. The identity of the viruses was confirmed via genome sequencing.

Funding: Pennsylvania Department of Agriculture.

PROJECT CONTACT
Diana Cox-Foster (dxc12@psu.edu)

Interactions of Israeli Acute Paralysis Virus (IAPV) and Deformed Wing Virus (DWV) on Honey Bee Health Examined

In a Cooperative Agricultural Project between ARS and the University of Illinois, scientists examined the interactions among IAPV, DWV, pesticides, and *Nosema*. Results showed that IAPV-infected bees exhibited 90–100 percent early (3–5 days post-infection) mortality rates and paralytic-like seizures before death. Behavioral changes included increased brood feedings and release of Nasanov pheromones, which draw bees back to the hive and are associated with worker recruitment to the hive or food and water sources. These results implicate the virus as being associated with both mortality and behavioral changes, which are possible links to CCD. The behavioral changes, in particular, deserve further investigation since CCD might be a consequence of such changes. Studies also showed that bees exposed to neonicotinoid insecticides and fungicides at sublethal levels had impaired immune systems and increased levels of the viruses noted above. Because CCD might be a result of such interactions and damaged bee immunity, further studies are planned to examine immune responses and determine the mechanisms underlying these interactions. Research is also characterizing levels of *Nosema ceranae* and *Nosema apis* to determine possible interactions and synergies with the viruses above.

Funding: USDA-NIFA and USDA-ARS.

PROJECT CONTACTS
Lee Solter (lsolter@illinois.edu)
Judy Chen (judy.chen@ars.usda.gov)
Jay Evans (jay.evans@ars.usda.gov)

Effect of Varroa Mites and Varroa-Transmitted Viruses on Nutritional and Health Status of Honey Bees

The main goal of this study (2010–2011) is to understand how mite infestation and mite-transmitted diseases affect immune competence and the nutritional status of managed honey bees. Preliminary results demonstrated that *Varroa* infestation has a major impact on the performance of individual honey bee health as well as on colony-level effects, including significant losses in weight in pupae and newly emerged adults infested with mites. Importantly, *Varroa* infestation affected the nutritional status of bees. The amount of available sugar, especially glucose, was depressed in *Varroa*-infested individuals, and this effect was significantly greater for adult bees than pupae. Interestingly, *Varroa*-infested bees were able to maintain their protein content, including protein content per unit weight in newly emerged bees bearing *Varroa* relative to controls. Consistent with the small effect of *Varroa* infestation on bee protein nutrition, the total free amino acid content did not vary significantly with treatment. The dominant amino acids in all samples (including all three life stages, and *Varroa*-infested vs. un-infested) were serine and proline. Preliminary results indicating that *Varroa*-infested pupae have reduced content of the key sulfur amino acid methionine are being investigated further, as are other metabolites. Additional efforts are focused on monitoring the level of viral infection and immune responses in *Varroa*-infested versus control bees using a DNA-based approach.

Funding: USDA-ARS and USDA-NIFA.

PROJECT CONTACTS

Katherine Aronstein (kate.aronstein@ars.usda.gov)
Angela E. Douglas (aes326@cornell.edu)

Viral Diseases and Varroa Mites Linked to Colony Losses

Since 2008, ARS scientists in Weslaco, Texas, have collaborated with university scientists to determine key factors (e.g., parasites, diseases, and pesticides) contributing to honey bee colony losses in apiaries across the United States. First-year data from this Coordinated Agricultural Project indicate that *Varroa* mite infestation and viral diseases contribute strongly to colony losses.

Funding: USDA-ARS and USDA-NIFA.

PROJECT CONTACT
Katherine Aronstein (kate.aronstein@ars.usda.gov)

Temperature Found to Affect Honey Bee Susceptibility to Fungi

Stonebrood and chalk brood are two fungal diseases of honey bees. Stonebrood is caused by a non-specialist fungal pathogen (*Aspergillus flavus*) that affects many animals to varying degrees. Chalkbrood is caused by a highly specialized pathogen of bees (i.e., *Ascosphaera apis).* When these fungal pathogens were fed to honey bees, stonebrood appeared to be more virulent than chalkbrood. However, when recently infected larvae were exposed to a 24 h cooling period (a cold stress), mortality from chalkbrood increased, whereas bees tended to recover from stonebrood. These results raise interesting questions about temperature stress and honey bee immune responses to pathogens with different degrees of specialization.

Funding: USDA-ARS and University of Copenhagen.

PROJECT CONTACTS
Rosalind James (rosalind.james@ars.usda.gov)
Svjeltlana Vojvodic (svo@life.ku.dk)

Temperature Stress Found to Increase Immune Response in Bees

Both high and low temperatures increased the expression of immune response genes in the alfalfa leafcutting bee. Genes associated with pathogen recognition and trypsin-like serine proteases were most highly expressed at the lowest rearing temperature (20°C), while prophenoloxidase, melanization, immune response signalling pathways, effectors, and reactive oxygen species were most highly expressed at the warmest temperature (35°C). The chalkbrood pathogen was found to affect some non-immunity host functions, particularly protein synthesis, differently at different temperatures, with the lowest rates of protein synthesis occurring at 30°C, a temperature where protein synthesis is high in healthy insects. If a temperature stress occurs before the pathogen has infected, it appears to elicit an immune response in the host, and this early response can prevent infection from occurring.

Funding: USDA-ARS.

PROJECT CONTACTS
Rosalind James (rosalind.james@ars.usda.gov)
Johan Hu (junhuan.xu@usu.edu)

Ongoing Research

Varroa Mite and Nosema Species Being Sequenced

ARS scientists in Beltsville, Maryland, have finished sequencing, annotating, and publishing the *Nosema ceranae* genome and have sequenced *Nosema apis*. Also, with industry support, ARS initiated a genome project on *Varroa destructor*, the primary honey bee parasite, as part of a worldwide consortium. Further analysis of these results will allow researchers to identify genes implicated in virulence, viral disease, and chemical resistance and to develop new strategies to control their impact on bees.

Funding: USDA-ARS and USDA-NIFA.
PROJECT CONTACTS
Judy Chen (judy.chen@ars.usda.gov)
Jay Evans (jay.evans@ars.usda.gov)

Topic IV: Mitigation and Management Preventive Measures

Goal 1: Develop best management practices for honey bees

1. Develop best management practices for migratory beekeeping.

Accomplishments

Best Management Practices (BMPs) Published

USDA-NIFA's Coordinated Agricultural Project (CAP) has published a Best Management Practices Guide for Honey Bee Health in the *American Bee Journal* and *Bee Culture*, which are two trade publications read by most beekeepers.

The BMP guide is also available on the CAP project's extension Web site, http://www.extension.org/pages/33379/best-management-practices-bmps-for-beekeepers-pollinating-californias-agricultural-crops. The BMP publication is the result of a collaborative effort involving CAP-funded scientists and Project Apis m., a beekeeper and almond grower foundation. For more information on the eXtension site, see Goal 7.

Funding: USDA-NIFA and Project Apis m.
PROJECT CONTACTS
John Skinner (jskinner@utk.edu)
Keith Delaplane (ksd@uga.edu)
Marion Ellis (mellis3@unl.edu)

Brood Pheromone Shown to Improve Colony Strength

ARS scientists in Weslaco, Texas, are testing a new brood pheromone device in honey bee colonies to improve the health of honey bees, as well as to improve crop pollination. In a cooperative agreement with Contech, Inc. (formerly PheroTech International Inc.), ARS carried out field trials showing that treated colonies experienced increases in pollen collection and population.

Funding: USDA-ARS and Contech, Inc.
PROJECT CONTACT
Katherine Aronstein (kate.aronstein@ars.usda.gov)

Effects of Supplemental Protein Feeding on Colony Growth Assessed

Cranberry pollination is extremely stressful to honey bees, providing minimal nutrition for the amount of work performed. ARS scientists studied whether supplemental feeding with the MegaBee® protein supplement during cranberry pollination could reduce colony losses and improve colony population growth. Results showed that colonies fed the protein supplement grew more than those that were not fed, indicating that even a relatively small addition of supplemental protein to colonies during cranberry pollination improves their growth and survival.

Funding: USDA-ARS and USDA-NIFA.
PROJECT CONTACTS
Gloria DeGrandi-Hoffman (Gloria.Hoffman@ars.usda.gov)
Gordon Wardell (GordyAW@aol.com)

Effects of Protein Supplements on Worker Physiology and Immune Response Examined

To evaluate the potential benefits of nutritional supplements on bee health, ARS scientists compared protein levels, endocrine development, and immune response in adult worker bees fed protein, pollen, and high fructose corn syrup supplements. Results showed that protein and pollen supplements produced similarly positive effects, but bees fed high fructose corn syrup had significantly reduced immune responses. This study will help beekeepers improve management strategies for their bees.

Funding: USDA-ARS.
PROJECT CONTACTS
Gloria DeGrandi-Hoffman (Gloria.Hoffman@ars.usda.gov)
Judy Chen (judy.chen@ars.usda.gov)

Winter Feeding and Its Improvement of Colony Size and Survival

This work started in the fall of 2006 and has continued to date. The Weslaco laboratory has shown that colonies being prepared for almond pollination in California show dramatic improvement when fed continuously from September through late January. Cooperating beekeepers and others have stated that they consider this work to have caused a total change in how they manage their colonies for almond pollination. Projects completed include interaction of feeding and *Nosema* infection; interaction of feeding and *Varroa* parasitism; optimum time of feeding; evaluation of commercially available pollen substitutes/ supplements; interaction of nutrition, *Nosema*, and *Varroa*; and mountainside wintering (cool, windy conditions reduce winter flight and the need for food); and the food efficiency (bees per pound of diet consumed) of four strains of bees. During feeding trials, some unanticipated findings came to light: a) colonies that went queen-less did not attempt supercedure or emergency queen cell construction (the duration of queen survival was correlated with the level of *Varroa* parasitism and to a lesser degree, the level of *Nosema* infection); b) colonies whose *Varroa* levels were above 150 mites (natural mite drop onto a sticky board during 3 days) in October had few colonies achieve almond pollination grade regardless of mite treatment or feeding in the fall-winter; c) treating colonies with fumagillin appears to have a short (about 2 months) negative impact on colonies. However, untreated infected colonies showed extremely poor performance after 14 months. This highlights the need to control even fairly moderate infections (500,000 spores/bee) of *N. ceranae* in some areas.

Funding: USDA-ARS.

PROJECT CONTACTS
Frank Eischen (frank.eischen@ars.usda.gov)
John Adamczyk (john.adamczyk@ars.usda.gov)

Honey Compound Could Enhance Bee Tolerance to Pesticide

Honey was found to contain a compound called pinocembrin, which is likely derived from propolis, and which acts as a powerful inducer of the enzymes that metabolize *tau*-fluvalinate. These results suggest that bee consumption of honey may serve a hitherto unrecognized role in honey bee health in the context of pesticide tolerance.

PROJECT CONTACT
May R. Berenbaum (maybe@illinois.edu)

Plant Resin Identified to Improve Bee Resistance to Fungal Toxins

Honey bees and their resource-rich nests are hosts to a wide range of fungi, including species that produce toxic substances called mycotoxins, which workers detoxify using gut enzymes. In bioassay experiments, bees that consumed extracts of propolis, a complex mixture of plant resins collected by bees and used as a general caulking material in the hive, had an enhanced capacity to detoxify mycotoxins. These results suggest that the enzymes involved in mycotoxin detoxification are enhanced by ingestion of propolis, which may serve a role in honey bee health.

PROJECT CONTACT
May R. Berenbaum (maybe@illinois.edu)

Malnutrition, Beneficial Microbes, and Nutritional Needs of Honey Bees

Scientists at ARS in Tucson, Arizona, believe that malnutrition is a major cause of colony losses. However, malnutrition, especially in its early stages, is difficult to diagnose. To remedy this, these ARS scientists are identifying biomarkers associated with the nutritional state of the bees.

Researchers found that bees emitted volatile compounds when adult workers and larval bees were malnourished. Emissions of these compounds increase in bees experiencing acute malnutrition. Identifying these chemical cues will enable ARS to evaluate how contemporary beekeeping and pollination practice, as well as environmental contamination, affects the nutritional state of a colony and its vulnerability to disease and decline.

Researchers at ARS in Tucson have uncovered key bacterial communities needed by bees for food processing and digestion. When scientists fed colonies pollen contaminated with fungicides, less than 30 percent reared new queens, suggesting that key nutrients needed to raise queens might be missing. Colonies that cannot successfully replace their queens cannot survive and thus contribute to the colony losses experienced by U.S. beekeepers.

Supplemental feeding can have deleterious effects on colonies if used for prolonged periods. ARS researchers showed that colonies fed supplemental protein produced higher numbers of brood but did not supply them complete nutrition. The lowered protein levels in workers shortened their lifespan and eventually reduced brood rearing, causing the colony to decline and the hives to perish. Additional studies found that during the winter, colonies fed sugar syrup made with sucrose produced more brood in spring compared with colonies fed high fructose corn syrup. These researchers are advising beekeepers that feeding colonies exclusively on high fructose corn syrup during winter or floral dearth periods can negatively impact colony strength required for pollination.

Funding: USDA-ARS.

PROJECT CONTACT
Gloria DeGrandi-Hoffman (Gloria.Hoffman@ars.usda.gov)

2. Develop best management practices for pest and pathogen control.

Accomplishments

High Ozone Concentrations Being Developed as a Fumigant for Beehive Equipment

In 2010, a mid-size ozone fumigation chamber was set up and field tested in Florida using honey bee supers and comb from beekeepers. Ozone fumigant killed 100 percent of the insect pests (small hive beetles and various wax moths), and pathogen loads were reduced. Pesticide residues were not effectively eliminated because comb wax protected the chemicals from degradation. Although the combs smelled unpleasant after treatment, no toxic compounds were formed, and the bees readily accepted ozone-treated comb in the colony. Treating comb also did not appear to affect the attractiveness to greater wax moth.

Funding: USDA-ARS, Florida Department of Plant Industries, and USDA-NIFA.

PROJECT CONTACT
Rosalind James (Rosalind.James@ars.usda.gov)

HopGuard®, a New Natural Product for Varroa Mite Control

Beekeepers need new methods to control *Varroa* because currently registered products are inconsistent in their effectiveness, harmful to brood, contaminate wax combs used in hives, or no longer control the mite because it is resistant. Under a Cooperative Research and Development Agreement (CRADA) with J.I. Haas, Inc., a product was developed (HopGuard®) that incorporates the beta plant acids (BPA) that are byproducts of hops processing for the beer brewing industry. Prototypes of the product were tested in commercial colonies and in packages of bees used to start new colonies. HopGuard® was highly effective in reducing mite populations. The product did not harm queens, adult bees, or brood, and it did not contaminate honey or the wax comb. A Section 18 emergency registration was issued by EPA and HopGuard®, and is now in commercial production.

Funding: CRADA between USDA-ARS and J.I. Haas, Inc.
PROJECT CONTACT
Gloria DeGrandi-Hoffman (Gloria.Hoffman@ars.usda.gov)

Controlling the Key Parasite Varroa destructor through Application of Genomics

A genome draft sequence was published for the *Varroa* mite, revealing potential weak points in mite biology (defensive proteins and proteins used in chemical mitigation) and candidates for novel controls such as RNA interference (RNAi). In fact, the publication in genomic databases of mite candidate genes allowed the worldwide initiation of RNAi efforts for this parasite, ending in late 2010, with the first successful demonstration of RNAi activity in *Varroa*. Microbes identified in this screen have also been screened across bees and mites as plausible controls for *Varroa*.

Funding: USDA-ARS.
PROJECT CONTACTS
Jay Evans (jay.evans@ars.usda.gov)
Scott Cornmann (scott.cornmann@ars.usda.gov)

Chemicals Associated with Mite Control Identified

ARS scientists at Gainesville, Florida, have discovered several chemicals that are produced in large amounts by drone and worker brood during the cell capping process but not by queen larvae, which affect the searching behavior of the *Varroa* mite.

ARS has filed a provisional patent on the use of these compounds for control of the *Varroa* mite and is working with private industry to develop control programs using the attractants.

Funding: USDA-ARS.
PROJECT CONTACT
Peter Teal (Peter.Teal@ars.usda.gov)

New Lures and Traps for Control of Small Hive Beetle (SHB) and Prevention of Invasion by Beetles Related to SHB

Working with scientists at the Department of Entomology and Nematology, University of Florida, ARS Gainesville scientists found that chemicals from bumble bee colonies were attractive to SHB. The beetle showed no preference for the honey bees over the bumble bees.

ARS scientists in Gainesville have developed a highly effective lure for the SHB based on chemicals produced by ripe fruit. The attractant has outperformed currently used attractants for the pests.

Also, scientists in Gainesville have developed a light-based attractant-trap system for capture of small hive beetle adults and larvae in honey houses. Tests show that the trap captures at least 80 percent of all wandering larvae.

These scientists also documented that the SHB reproduces as well on ripe fruit as it does in honey bee colonies. This finding, as well as the fact that beetles are more strongly attracted to odors from fruit like pears and cantaloupe, demonstrates that the pest is very capable of surviving in the absence of managed bee hives and provides important information for developing strategies to mitigate beetle invasion of hives.

Scientists at Gainesville, working with scientists from the International Center for Insect Physiology and Ecology in Nairobi, Kenya, have also identified a new beetle pest of honey bee colonies in Africa and have discovered that the beetle is attracted to lures currently in use to monitor for the SHB.

These discoveries are important because we can now monitor for this pest at ports of entry into the United States and thereby know if and when the new beetle invades the United States.

Funding: USDA-ARS.
PROJECT CONTACT
Peter Teal (Peter.Teal@ars.usda.gov)

Major Bee Pathogen Sequenced

Recently, ARS scientists in Weslaco, Texas, finished sequencing one of the major honey bee pathogens, *Ascosphaera apis*, the causative agent of chalkbrood disease. Research focuses on discovering genes involved in the invasion and pathogenesis of the bee host. This information will lead to the development of novel disease control methods.

Funding: USDA-ARS.
PROJECT CONTACT
Katherine Aronstein (kate.aronstein@ars.usda.gov)

Chalkbrood Disease and Implementations to Host Pathogenesis

ARS scientists analyzed responses of the honey bee fungal pathogen *Ascosphaera apis* during host pathogenesis using high-throughput sequencing. Data analysis captured a significant number of differentially expressed genes that may be related to fungal reproduction, host invasion, and stress responses.

This study may provide new insights into the mechanisms of host invasion and progression of chalkbrood disease and presents a unique opportunity to improve our understanding of the highly complex nature of host-pathogen interactions.

Funding: USDA-ARS.
PROJECT CONTACTS
Katherine Aronstein (kate.aronstein@ars.usda.gov)
Jay Evans (jay.evans@ars.usda.gov)
Scott Cornmann (scott.cornmann@ars.usda.gov)
Anna Bennett (akb39@hoyamail.georgtown.edu)

3. Establish guidelines for floral gardens to maintain stronger honey bees. and
4. Develop best management practices for pesticide use.

Accomplishment

Progress Toward Best Management Practices for Pesticide Use

EPA is working with a broad range of stakeholders through the Pesticide Program Dialogue Committee (PPDC), a stakeholder group chartered under the Federal Advisory

Committee Act, to explore and develop additional ways of mitigating the potential effects of pesticides on bees. Broad stakeholder input will best ensure success of best management practices (BMPs) that may involve not only beekeepers, but also growers and pesticide applicators. Additional mitigation measures under consideration include both improved label language and encouraging communication/cooperation between growers/applicators and beekeepers.

EPA has been working through its regional offices and with State Lead Agencies to engage local stakeholders in dialogue to understand successful grower-beekeeper interactions on a local scale. Understanding successes at the local level may serve as a basis to expand or transplant local successes to other State or national programs. Programs like Driftwatch™ (http://driftwatch.agriculture.purdue.edu/index.html) are being examined as a means of identifying the location of vulnerable "crops" like honey bee colonies (apiaries) through national registries that would enable growers/applicators to be aware of such vulnerable areas in advance of pesticide applications and communicate pesticide application options with the beekeepers before they occur.

As discussed earlier, EPA is also working with its OECD international partners to develop a compendium of mitigation measures that various countries have used around the world to mitigate the potential risks of pesticides to insect pollinators.

PROJECT CONTACTS
Tom Steeger (Steeger.Thomas@epa.gov)
Tom Moriarty (Moriarty.Thomas@epamail.epa.gov)

Goal 2: Develop Best Management Practices for Non-Apis Bees to Provide Alternative Pollinators for Crops, Gardens, and Natural Areas

1. Develop best management practices for pest and pathogen control in non-*Apis* bees.

and

2. Establish guidelines for maintaining stronger populations of non-*Apis* bees in agricultural systems, home gardens, and wild lands.

Accomplishments

Blue Orchard Bee Workshops Delivered to Almond Production Community

ARS scientists with California Cooperative Extension personnel organized and delivered workshops (2009 and 2010) to provide information on the blue orchard bee and its use for almond pollination. Workshops were designed to facilitate networking among the almond production community interested in using blue orchard bees as almond pollinators along with, or instead of, honey bees. Guidance on managing blue orchard bees for almonds is now available to promote its use and reduce pressure on the sole use of honey bees.

Funding: USDA-ARS and University of California–Davis.
PROJECT CONTACTS
James Cane (jim.cane@ars.usda.gov)
Theresa Pitts-Singer (theresa.pitts-singer@ars.usda.gov)
Carolyn Pickel (cxpickel@ucdavis.edu)

Efficiency of Alfalfa Leafcutting Bees Affected by Stocking Density and Floral Resource Abundance

The alfalfa leafcutting bee (*M. rotundata*) is commonly used to pollinate alfalfa for seed production in the United States and Canada, but maintaining healthy stocks of these bees is hampered by disease, parasites, predators, and unexplained mortality of eggs and small larvae. An ARS researcher has shown that typical U.S. alfalfa leafcutting bee stocking densities for alfalfa pollination are unnecessarily high, decreasing the number of bees that remain at the commercial site, decreasing bee pollination efficiency, and reducing the number of healthy offspring produced. The amount of alfalfa bloom in fields also affects the number of bees that remain onsite and the number of offspring that can be produced. This research provides information for optimizing stocking density in light of available floral resources.

Funding: USDA-ARS.

PROJECT CONTACT
Theresa Pitts-Singer (theresa.pitts-singer@ars.usda.gov)

Delaying Novaluron Applications Decreases Negative Effects

Novaluron (Rimon®) is a relatively new insecticide that is an insect growth regulator, labeled for lygus control in alfalfa fields during bloom. Some alfalfa leafcutting beekeepers have complained about poor bee returns in fields where novaluron was used, and laboratory trials have shown this pesticide to be toxic to larvae and eggs. Field experiments were conducted in 2010 to evaluate the effect of the timing of novaluron (Rimon® 0.83 EC) spray applications on alfalfa leafcutting bee reproduction. Spray timing was found to have no impact on the mean number of nest cells produced by the end of the season. However, the percentage of pollen balls (bees that died as eggs or young larvae) was lower, and the percentage of live larvae higher, on unsprayed (control) plots and plots treated with Rimon® late, as compared to plots treated during early and mid-bloom. These results indicate that Rimon® has a reduced impact on overall production of healthy alfalfa leafcutting bees when applications are delayed until the late bloom period.

Funding: USDA-ARS, USDA-NIFA, and Chemtura.

PROJECT CONTACTS
Theresa Pitts-Singer (theresa.pitts-singer@ars.usda.gov)
James Barbour (jbarbour@uidho.edu)

Nesting Methods Successfully Established for Four Bumble Bee Species

Bumble bees are important pollinators of commercial greenhouse crops, but they are difficult to raise in captivity. Because it is particularly difficult to get queens to establish new nests in captivity, three nest establishment methods were tested on four species of bumble bees. The results show that some techniques greatly increase the success rate (by fourfold) in some species in comparison to other techniques. By targeting the proper rearing techniques to a given species, bumble bee producers and researchers will be able to save time and resources when producing these bees in culture.

Funding: USDA-ARS.

PROJECT CONTACT

James Strange (james.strange@ars.usda.gov)

Wildland Fire Shown to Increase Wild Pollinators

Controlled burns constitute one method used to manage fuel loads in natural areas, but the effects on pollinators remain unknown. The fates of wild bee communities following wildfire were largely unknown but are critical to anticipating pollination services during large post-fire rangeland restoration projects. Using a new suite of sampling protocols, researchers systematically sampled bee floral guilds over time after large burns, sampling from sites inside and outside the burned areas. Plant communities were also characterized. Where rangeland plant communities were in good shape before burning, the diversity of the native bee communities returned the following year. Over several years, the native wildflowers likewise recovered. Initial seeding in restoration projects should include plants that are broadly attractive to local native bees to sustain these bees until the native plant communities can reestablish, since that re-establishment can take several years.

Funding: USDA-ARS, DOI-BLM, and USDA-FS.

PROJECT CONTACT
James Cane (jim.cane@ars.usda.gov)

Goal 3: Maintain Bees with Resistance to Parasites and Pathogens

1. Identify traits associated with resistance to parasites and pathogens.

Accomplishments

USDA Varroa-Resistant Honey Bees Found Well Suited for Commercial Pollination in Two Trials

ARS scientists tested Russian honey bees and bees containing a *Varroa*-resistant trait (*Varroa*-sensitive hygiene [VSH]) for size and productivity in two large-scale field tests comprising two different migratory routes. Two years of testing during commercial pollination of almonds, apples, blueberries, and cranberries, with overwintering in Louisiana, showed the USDA *Varroa*-resistant stocks to be as large and productive as control stocks at each pollination site. One year of a test involving almond pollination and Midwest honey production, with overwintering in California's Central Valley, again showed Russian and VSH bees to have adequate colony size, survivability, and honey production. These results suggest that *Varroa*resistant bees are well-suited for commercial pollination.

Funding: USDA-ARS.

PROJECT CONTACTS
Tom Rinderer (Tom.Rinderer@ars.usda.gov)
Bob Danka (Bob.Danka@ars.usda.gov)

Improved Management Strategies for Russian Honey Bees Developed

ARS scientists at Baton Rouge, Louisiana, have evaluated management procedures to improve Russian honey bees for early season pollination of almonds. Extensive tests found that feeding regimes (and secondarily, hive size) affect colony populations. Colonies fed pollen substitute dwindled less in winter than those not fed. Feeding pollen substitute with natural pollen added was much better than feeding just pollen substitute. Feeding protein as patties was better than feeding it in plastic comb within the brood nest. Feeding sucrose syrup at a slow rate continuously from autumn to mid-winter simultaneously with protein increased colony size. Russian colonies grew more in 8-frame hives than in 10-frame hives, but hive size was less important than feeding protein. Housing colonies in black hives did not improve colony population over colonies in white hives.

These findings collectively show that Russian honey bees can be managed in the southern United States to produce colonies that are more than sufficient to rent for almond pollination in February.

Funding: USDA-ARS.

PROJECT CONTACTS
Tom Rinderer (Tom.Rinderer@ars.usda.gov)
Bob Danka (Bob.Danka@ars.usda.gov)

Ongoing Research

Identifying Genes that Confer Resistance to Varroa Mites

Two separate projects were undertaken, and genes were mapped to chromosomal regions that influence each of the two behavioral traits that have been shown to be important for suppressing mite populations: *Varroa*-sensitive hygiene (VSH) and mite-grooming behavior. Several chromosomal regions were identified for each trait.

The genes in these regions are being analyzed and considered for follow-up experiments to confirm their role in conferring resistance to *Varroa* so that they can be targeted with marker-assisted selection to pyramid multiple resistance genes in commercial stocks.

Funding: USDA-NIFA and USDA-ARS.

PROJECT CONTACTS
Greg Hunt (ghunt@purdue.edu)
Marla Spivak (spiva001@umn.edu)
Bob Danka (Bob.Danka@ars.usda.gov)
Miguel Arechavaleta-Velasco (are@servidor.unam.mx)
Thomas Webster (thomas.webster@kysu.edu)
Katherine Aronstein (kate.aronstein@ars.usda.gov)

Also see Goal 4, Objective 1: Identifying genes that confer resistance to Varroa *and pathogens, and genes that respond to biotic challenges.*

2. Introduce resistance traits into bee stocks favored by the industry.

Accomplishments

Breeding for Hygienic Behavior

University of Minnesota researchers and extension specialists are evaluating commercial honey bee stocks for the level of hygienic behavior, a trait that is associated with the bee's resistance to the *Varroa* mite. A pilot certification program will be initiated for the hygienic trait. A "Tech Transfer" team is forming in California to assist bee breeders with stock selection for hygienic behavior and to reduce chemical use in breeder colonies. It is envisioned that this team will serve as a model for the Nation in producing local, regionally adapted queens that are resistant to mites.

Funding: National Honey Board and California Almond Board.

PROJECT CONTACT
Marla Spivak (spiva001@umn.edu)

Commercialization of Varroa-resistant Honey Bees Selected for Pollination Performance

Varroa mites are a major cause of colony losses throughout the United States. Bees with *Varroa*-sensitive hygiene (VSH), which have good resistance to *Varroa* mites, were tested by researchers for two seasons in a commercial migratory beekeeping operation. Bee colonies were created from VSH queens that were out-crossed, i.e., matings were not controlled, a method used by most large-scale beekeepers. Bee colonies were shipped nationwide for pollination of four crops and for late-summer honey production. VSH colonies performed well in terms of survival, populations, and resistance to *Varroa* mites. The best surviving VSH bee colonies from each year were propagated to form a VSH breeding population that had enhanced genetics for both mite resistance and behavior related to crop pollination. These bees are now being marketed by a Cooperative Research and Development Agreement partner, Glenn Apiaries, and the use of their germplasm should improve adoption of mite-resistant bees by commercial beekeepers that pollinate crops.

Funding: USDA-ARS.

PROJECT CONTACTS
Bob Danka (Bob.Danka@ars.usda.gov)
Tom Rinderer (Tom.Rinderer@ars.usda.gov)

Ongoing Work

Genetic Techniques Used to Improve Bee Breeding

The University of Tennessee has replaced half its bee stock with varieties resistant to *Varroa* mites (either the *Varroa*-Sensitive Hygiene [VSH] strain or with hybrids raised from VSH queens). Equipment and apiary expansion is underway to conduct bee breeding demonstration workshops in 2011 and 2012.

Workshop materials will be obtained or developed for inclusion on eXtension.org for use in any bee breeding workshop. These efforts will increase the use of robust bee stocks.

Funding: USDA-NIFA.

PROJECT CONTACTS
Nick Calderone (nwc4@cornell.edu)
Keith Delaplane (ksd@uga.edu)
Greg Hunt (ghunt@purdue.edu)
Steve Sheppard (shepp@wsu.edu)
Marla Spivak (spiva001@umn.edu)
Thomas Webster (thomas.webster@kysu.edu)

Michael Wilson (mwilso14@utk.edu)

3. Use genomic technologies and germplasm preservation to maintain quantities of desirable honey bee germplasm.

Accomplishments

Honey Bee Germplasm Importations to Improve Genetic Variation in United States

Honey bee semen was collected from locations in Italy and the Republic of Georgia in 2010 and returned to the United States under a USDA-APHIS quarantine permit to Washington State University. The fresh semen was used to inseminate queens produced from existing U.S. commercial strains of honey bees and overwintered in Washington and California (collaboration with University of California–Davis). This genetic material is being incorporated into existing U.S, honey bee strains to increase sex allele diversity and reduce inbreeding effects. Genetic material derived from the Italian importations (2008–2010) will be released to the commercial bee breeding industry in 2011.

Funding: Washington and California State Beekeeping Associations and USDA-NIFA-AFRI-CAP.

PROJECT CONTACT
Steve Sheppard (shepp@wsu.edu)

Cryopreservation of Honey Bee Germplasm

In 2010, the Washington State University research team also developed a practical methodology for maintaining honey bee germplasm (semen) by cooling samples at temperatures below the freezing point. Using this new method, researchers there produced three successive generations of honey bee queens from cryopreserved (i.e., frozen at liquid nitrogen temperatures) semen in 2010. This demonstration marked the turning point for a new era of germplasm preservation/conservation in honey bees, as it will now be possible to establish genetic repositories for the honey bee. Future germplasm collections (2011 and beyond) will include cryopreservation of a portion of the semen for subsequent use. A manuscript for publication in a peer-reviewed journal was prepared and submitted on this topic.

Funding: Washington and California State Beekeeping Associations and USDA-NIFA-AFRI-CAP.

PROJECT CONTACT
Steve Sheppard (shepp@wsu.edu)

3. Transition to mite- and pathogen-resistant honey bee stocks.

Accomplishments

Groundwork Established for a Sustainable Market for Genetically Improved Queens

Researchers are working to train beekeepers in stock selection (e.g., VSH bees) and breeding techniques, as well as management and testing practices to reduce impact from pests, pathogens, and chemicals. Specifically, scientists are teaching a 2-day Queen Rearing Short Course every summer at the University of Minnesota to assist bee breeders. This course combines lecture and hands-on field work.

In 2008 and 2009, researchers conducted onsite visits with 17 queen producers in northern California. These operations produce at least half of the commercially available queens in the Nation and thus have major impact on honey bee genetics and improvement nationally. Scientists demonstrated how to test their stock for hygienic behavior and how to test their colonies for *Nosema*, and they provided best management practices, including reducing chemical inputs into their breeding stocks.

Washington State University and University of California at Davis researchers have imported selected stock for two Carniolan and Italian bees for breeding purposes. In the future, the team will bring in a third subspecies favored by beekeepers, the Caucasian.

Over the 2-year study, two things became evident: (1) bee breeders are very competent and produce high-quality, well-mated queens to supply beekeepers throughout the Nation; (2) breeding for pathogenic resistance is highly complex, while maintaining productive characteristics and pollination efficiency requires professional assistance to help bee breeders improve stock selection, enhance genetic diversity, and perform disease diagnostics.

Funding: USDA-NIFA.

PROJECT CONTACTS

Keith Delaplane (ksd@uga.edu)
Marla Spivak (spiva001@umn.edu)

Increased Transfer of the VSH Mite-Resistance Technology to Foster Breeding of Bees with Improved Mite Resistance

Bees with the trait of *Varroa*-sensitive hygiene (VSH) have strong resistance to *Varroa* mites. VSH breeder queens are available through an ARS partnership with a commercial queen breeder. Semen from VSH bees now is directly available from ARS for incorporation into existing bee stocks. Research showed that commercially available VSH queens that are mated to non-VSH drones have acceptable mite resistance. Tests are ongoing in Hawaii to determine the value of VSH in mitigating *Varroa* damage in a tropical environment that is the base of a large queen production industry. Two workshops were given in 2010 at regional and national beekeeping meetings to teach breeders how to test, select, and breed for the trait. Together, these efforts should encourage incorporation of VSH-based mite resistance into diverse bee stocks and thus support beekeeping sustainability.

Funding: USDA-ARS.

PROJECT CONTACTS

Tom Rinderer (Tom.Rinderer@ars.usda.gov)
Bob Danka (Bob.Danka@ars.usda.gov)

Also see Topic IV, Goal 3, Objective 2.

Goal 4: Develop Ways to Manage Mite Resistance to Miticides and Create Alternatives

1. Develop resistance management programs that provide beekeepers with tools for mite management.

Accomplishment

EPA Expediting Reviews for Section 18

EPA is working with State Lead Agencies to provide beekeepers with appropriate tools to control bee colony pests such as *Varroa* mites. The Agency is working closely with USDA ARS and the Office of Pest Management Policy (OPMP) to expedite reviews of Emergency Exemption requests (under section 18 of the Federal Insecticide, Fungicide and Rodenticide Act [FIFRA]), Special Local Needs Registrations (under section 24c of FIFRA), and full registrations (under section 3 of FIFRA) of miticides.

PROJECT CONTACTS
Tom Steeger (Steeger.Thomas@epa.gov)
Tom Moriarty (Moriarty.Thomas@epamail.epa.gov)

1. Develop new methods of managing parasites and pathogens. See Type IV, Goal 3.

Goal 5: Improve the Regulatory Framework to Better Protect Against the Introduction of New Pathogens, Pests, and Parasites of Bees to Meet World Trade Organization (WTO) and International Committee of the World Organization for Animal Health (OIE) Requirements for the Importation and Exportation of Honey Bees

1. Develop new molecular detection systems that can be used to detect pathogens, pests, and parasites in introduced bee stocks and bee products used in beekeeping.

and

2. Explore opportunities to change regulations based on new molecular detection systems. For example, if a virus is shown to be pathogenic and reliable new detection methods are found, then movement of virus-contaminate bee stocks may be regulated by bodies such as OIE.

Accomplishments

New Molecular Assay for Resistant Bacterial Strains Developed

ARS identified the mechanism of resistance of the bacterial honey bee pathogen (for American foulbrood) to the antibiotics oxytetracycline (OTC) and developed a new molecular assay for detecting OTC-resistant bacterial strains in bee samples. This new identification method is now available for regulatory officials who are considering ways of preventing the spread of resistant bacterial strains in bee shipments.

Funding: USDA-ARS.
PROJECT CONTACT
Katherine Aronstein (kate.aronstein@ars.usda.gov)

Guidelines for Bee Imports Developed

The North American Plant Protection Organization completed recommendations to coordinate regulation and importations of non-*Apis* bees between Mexico, Canada, and the United States. Initial guidelines were developed for the safe importation of bees to avoid the accidental release of bee pathogens and parasites and to avoid the introduction of bee species that may cause a loss of native bees.

Funding: USDA-APHIS and USDA-ARS.
PROJECT CONTACTS
Colin Stewart (Colin.Stewart@aphis.usda.gov)
Rosalind R. James (Rjames@biology.usu.du)

1. Establish processes for periodic monitoring of the U.S. honey bee population to determine whether specific pests are present.

Accomplishment

Pilot Survey for Bee Diseases, Parasites, and Pests

APHIS, in collaboration with ARS, conducted a pilot survey in 2009 in Florida, California, and Hawaii to validate sampling protocols used to determine which diseases/parasites/pests of honey bees are and are not present in the United States. A limited national survey of 13 States was conducted in 2010. A report of these results can be found at http://www.aphis.usda.gov /plant health/plant pest info/honey bees/downloads/2010-2011-Limited_ Survey_Report.pdf.

A national survey is being conducted in 2011 in 34 States. The exotic pests Tropilaelaps mites, *Apis ceranae*, and Slow Paralysis Virus have not been found during this survey. Therefore, the United States is not allowing importations from countries known to have these pests (e.g., Australia and Southeast Asia).

Funding: USDA-APHIS and USDA-ARS.
PROJECT CONTACTS
Robyn Rose (Robyn.I.Rose@aphis.usda.gov)
Jeff Pettis (jeff.pettis@ars.usda.gov)

Goal 6: Demonstrate Improved Colony Health by Integrating Research-Derived Knowledge and Tactics into an Area-Wide Project

1. Test and verify management approaches for mite control, improved diet, improved bee stock, and changes in migratory practice.

and

2. Transfer technology for early spring bee availability for pollination.

Accomplishment

Area-wide Project Initiated

ARS is continuing an Area-wide Project on Honey Bee Health across multiple ARS locations (Tucson, Arizona; Beltsville, Maryland; Weslaco, Texas; and Baton Rouge, Louisiana). The project has several aspects, including documenting the impact of migration on bee colonies, examining the effects of supplemental feeding on colony health, developing more resistant bee lines and improved methods for their management, and developing better control methods for honey bee pests. Ultimately, the project aims to develop a set of best management practices for migratory beekeepers to reduce stress on their bee colonies, thereby enabling bees to ward off threats. Specific results of these individual research projects are described elsewhere in this report.

Funding: USDA-ARS.

PROJECT CONTACTS
Jeff Pettis (jeff.pettis@ars.usda.gov)
Tom Rinderer (Tom.Rinderer@ars.usda.gov)
John Adamczyk (John.Adamczyk@ars.usda.gov)
Frank Eischen (Feischen@weslaco.ars.usda.gov)
Gloria DeGrandi-Hoffman (Gloria.Hoffman@ars.usda.gov)

Also see Topic III, and Topic IV, Goal 1.

Goal 7: Transmit or Disseminate Science-Based Information to Manage Bees

1. Develop, maintain, and preserve a secure Web-based site for scientific collaboration (SharePoint).

and

2. Develop, maintain, and update a Web-based public Internet site (e.g., eXtension or PIPE; Pest Information Platform for Extension)

Accomplishments

eXtension Progress

- The eXtension.org Community of Practice (CoP) continues to disseminate Web-based information on bee health (http://www.extension.org/bee_health). Currently, the Bee Health CoP has over 100 contributors with expertise in bee research and extension. There are approximately 400 pages of content on this Web site. A few examples include (1) The Best Management Practices For Beekeepers Pollinating California's Agricultural Crops, (2) *Varroa*-Sensitive Hygiene and Mite Reproduction, and (3) American Bee Research Conference proceedings with videos of scientific presentations. An average of 11,000 page views per month was recorded during a 12-week period in 2010. Also, the CoP features an "Ask an Expert" tool that allows the public to ask bee health professionals questions regarding concerns about bees. Users can upload pictures to facilitate better disease and pest identification. To date, about 500 questions about bees were successfully answered by the experts.
- A new team of 27 bee research and extension specialists has organized around the CoP to expand on surveys of honey bee losses (such as the winter loss survey) and sampling projects (such as the California Bee Breeders stock improvement effort). A new Web site was created, which solicits survey participation and provides information about their efforts (http://beeinformed.org/). A blog page was also created to foster informal discussions about bee-related topics between project cooperators and the public (http://beeinformed.org/blog).
- Other ongoing Web-related outreach tools include email campaigns, articles in beekeeping journals with an online presence (e.g., Bee Culture and American Bee Journal), and results from research by NIFA's Managed Pollinator CAP team (http://www.beeccdcap. uga.edu).

Funding: USDA-NIFA and USDA-ARS.

PROJECT CONTACTS

Keith Delaplane (Ksd@uga.edu)
John A. Skinner (Jskinner@utk.edu)
Zachary Huang (Huang@msu.edu)
Greg Hunt (GHunt@purdue.edu)
May Berenbaum (Maybe@uiuc.edu)

Social Media Outreach Initiated

Researchers are pursuing additional outreach opportunities using social networking sites and public Web pages. Honey bee health is featured on a YouTube page (http://www.youtube.com/beehealth), and researchers are now implementing pages on Facebook, Twitter, Second Life, and Wikipedia, and beekeeper Web forums such as www.beesource.com.

Funding: USDA-NIFA.

PROJECT CONTACTS

John A. Skinner (Jskinner@utk.edu)
Keith Delaplane (Ksd@uga.edu)

End Notes

[1] The first Pellston Conference was held in 1977 to address the needs and means for assessing the hazards of chemicals to aquatic life. Since then, many conferences have been held to evaluate current and prospective environmental issues. Each has focused on a relevant environmental topic, and the proceedings of each have been published as a peer-reviewed or informal report. These documents have been widely distributed and are valued by environmental scientists, engineers, regulators, and managers because of their technical basis and their comprehensive, state-of-the-science reviews. The first four Pellston conferences were initiated before SETAC was effectively functioning. Beginning with the 1982 conference, however, SETAC has been the primary organizer, and SETAC members (on a volunteer basis) have been instrumental in planning, conducting, and disseminating conference results. Taken from: http://www.setac.org/node/104

[2]OECD 2010. Environmental Directorate Joint Meeting on the Chemicals Committee and the Working Party on Chemicals, Pesticides and Biotechnology (ENV/JM/MONO(2010)24 Series on Pesticides No. 52 OECD Survey on Pollinator Testing, Research, Mitigation and Information Management: Survey Results http://www.oecd.org/dataoecd/19/27 /45275778.pdf.

In: Bee Health: Factors, Analyses, and Research Progress ISBN: 978-1-62618-044-4
Editor: Roger Dunham

Chapter 7

COLONY COLLAPSE DISORDER PROGRESS REPORT, JUNE 2010*

CCD Steering Committee

EXECUTIVE SUMMARY

Mandated by the 2008 Farm Bill [Section 7204 (h) (4)], this second annual report on Honey Bee Colony Collapse Disorder (CCD) research progress represents the work of a large number of scientists from 8 Federal agencies, 2 state departments of agriculture, 22 universities, and several private research efforts.

In response to the unexplained losses of U.S. honey bee colonies, now known as colony collapse disorder (CCD), the U.S. Department of Agriculture's (USDA) Agricultural Research Service (ARS) and National Institute of Food and Agriculture (NIFA) led a collaborative effort to define an approach to CCD, resulting in the CCD Action Plan in July 2007.

Many universities and organizations—Federal, State, and private—were involved in developing this plan and are carrying out the work that addresses the CCD problem. The CCD Action Plan is organized under four topic areas:

1. Survey and (sample) data collection;
2. Analysis of existing samples;
3. Research to identify factors affecting honey bee health, including attempts to recreate CCD symptomology; and
4. Mitigation and preventive measures.

Highlights of progress in these topic areas are summarized below.

---

* This is an edited, reformatted and augmented version of report by the USDA Agricultural Research Service, dated June 2010.

Topic I: Survey and (Sample) Data Collection

Several survey and data collection efforts have been underway to provide baseline information on CCD, resulting in better defined CCD symptoms, and documented evidence of increasing honey bee losses. Colonies with poor health, including colonies with CCD symptoms, exhibited increased pathogen levels and evidence of pesticide residues. General colony loss levels in 2009 were 29 percent and increased to 34 percent in 2010. Losses are a major problem for the beekeeping industry and potentially to commodities that depend on bees for pollination services.

Topic II: Analysis of Existing Samples

Based on an initial analysis of collected bee samples (CCD- and non-CCD affected), reports have noted the high number of viruses and other pathogens, pesticides, and parasites present in CCD colonies, and lower levels in non-CCD colonies. This work suggests that a combination of environmental stressors may set off a cascade of events and contribute to a colony where weakened worker bees are more susceptible to pests and pathogens. Testing this complex hypothesis is the focus of joint efforts between Federal and university scientists funded by ARS and NIFA.

Topic III: Research to Identify Factors Affecting Honey Bee Health, Including Attempts to Recreate CCD Symptomology

Survey and analysis results indicate that many pesticides, parasites, and pathogens are associated with CCD. Therefore, efforts to explore the role of these factors continue. Findings currently suggest an association of sub-lethal effects of pesticides with CCD. Two common miticides in particular, coumaphos and fluvalinate, which are pesticides registered for use in bee colonies to control varroa mites, are suspect, either acting individually or in combination (e.g., synergistically, where the combination of the two compounds is more toxic than either compound alone). The emerging evidence of pesticide exposure to pollinators and potential interactive effects indicates the need to further study pesticides for their potential interactions with CCD. Studies have also confirmed suspected links between CCD and poor colony health, inadequate diet, and long-distance transportation. Research results indicate that both supplemental protein diets and natural pollen feedings can increase colony strength and offset the negative impacts of stress caused by pests, pesticides, and long-distance transportation of bees by beekeeping operations.

Topic IV: Mitigation and Management Measures

Efforts to mitigate honey bee losses are highlighted by two national multi-year projects, the ARS Areawide Project on Honey Bee Health and a NIFA-funded Coordinated Agricultural Project (CAP), which are allowing researchers to test various hypotheses related to poor bee health and provide further insights into the problem, and strategies for better bee management. Key accomplishments to date include the development of new, varroa mite-resistant bee stocks, improved diets and timing of feeding, a strategy (comb irradiation and ozone treatment) to reduce pathogen levels, and identification of several alternative pollinator species. Progress has also been made toward developing new detection capabilities for pests and pathogens, integrated control strategies for varroa mites, and a set of comprehensive best management practices that will provide beekeepers with specific guidelines to improve the health of their bees. As a method for potentially mitigating the occurrence of CCD, and in an effort to encourage the provision of improved and sustainable habitat for both managed honey bees and all pollinators, the Natural Resources Conservation Service (NRCS) is implementing its NRCS Pollinator Initiative, which is a multifaceted initiative involving the following: (1) the revision of conservation practice standards to better address pollinator habitat needs; (2) the identification and provision of recommended ecoregion-specific, pollinator-friendly vegetation as part of the implementation of the NRCS Plant Materials Centers' Pollinator Conservation Action Plan; (3) the provision of support and encouragement for enhancing pollinator habitat under a variety of conservation programs as identified in the NRCS Programs Pollinator Action Plan; and (4) the provision of awareness-building materials, training, and outreach about the importance of pollinators and helpful actions individuals may pursue. During the past 3 years, numerous causes for CCD have been proposed and investigated. Although the cause of CCD is still unknown, research has provided support to the hypothesis that CCD may be a syndrome caused by many different factors that work individually or in combination. Looking ahead, studies will focus increasingly on the interaction of multiple factors in causing CCD. More detailed findings follow in the body of the report.

Colony Collapse Disorder Annual Progress Report

This report is the second annual report prepared in response to Section section 1672 (h) (4) of the Food, Agriculture, Conservation, and Trade Act of 1990 (7 U.S.C. 5925 (h) (4)), as added by section 7204 (a) (3) of the Food, Conservation, and Energy Act of 2008 ("2008 Farm Bill") (Pub. L. 110-246), which directed the Secretary of Agriculture to:

> "submit to the Committee on Agriculture of the House of Representatives and the Committee on Agriculture, Nutrition, and Forestry of the Senate an annual report describing the progress made by the Department of Agriculture in—
>
> (A) investigating the cause or causes of honey bee colony collapse; and
> (B) finding appropriate strategies to reduce colony loss."

Background

After the large-scale, unexplained losses of managed U.S. honey bee (*Apis mellifera* L.) colonies during the winter of 2006-2007, investigators identified a set of symptoms that were termed colony collapse disorder (CCD). In response to this problem, Federal and State government, university, and private researchers, led by the U.S. Department of Agriculture's (USDA) Agricultural Research Service (ARS) and Cooperative State Research, Education, and Extension Service (now the National Institute of Food and Agriculture (NIFA)), mobilized to define an approach to CCD, an effort resulting in formation of the CCD Steering Committee, and publication of the CCD Action Plan in July 2007. Many organizations, public and private, in addition to those represented on the Steering Committee, are involved in the work to address the CCD problem.

During the past 3 years, numerous causes for CCD have been proposed and examined. There have been many associations identified throughout the course of research; however, it is becoming increasingly clear that no single factor alone is responsible for the malady. Researchers continue to document elevated pathogen levels in CCD-affected bees, with no specific pathogen linked definitively to CCD. In addition, tests to examine hives for known honey bee parasites (varroa mites, honey bee tracheal mites, Nosema species), which pose significant problems for beekeepers, and once were highly suspected to play a major role in CCD, have not revealed these parasites, by themselves, at sufficient levels to explain the problem. Over the past year, several independent studies have shown that bees are exposed to a wide range of pesticides and that some pesticides have interactive effects (on bee mortality) with other pesticides, with bee pests, or with viruses. Taken together, these studies support the hypothesis that CCD is a syndrome of stress, caused by many different factors working individually, but more likely in combination. Insufficient data are available to confirm this, but studies continue, based on the groundwork that has been laid.

Although the causes of CCD have not been fully illuminated, the research response is coordinated and dedicated to resolving this issue, while simultaneously improving pollinator health. NIFA's Coordinated Agricultural Project and ARS' Areawide Project on Honey Bee Health are of particular importance for their ongoing efforts to examine bee effects from a broad, regional perspective, which will be a key component to solving the crisis. Funding from ARS and NIFA, with additional contributions by a number of other sources, including the National Honey Board, the Almond Board of California, Burt's Bees, Häagen-Dazs, the North American Pollinator Protection Campaign, Project "Apis m." (PAm), the Foundation for the Preservation of Honey Bees and others, is resulting in a variety of new studies and new expertise on bee health issues. The fruits of this increased research effort are being published, and a new "eXtension" website (http://www.extension.org/bee%20health) has been assembled to provide reliable research-based information to beekeepers and the general public.

The CCD Action Plan is organized under four topic areas:

1. Survey and (sample) data collection;
2. Analysis of existing samples;
3. Research to identify factors affecting honey bee health, including attempts to recreate CCD symptomology; and
4. Mitigation and preventive measures.

Following up on the 2007-2008 Report, below is a summary of progress in meeting Action Plan objectives for each of these topic areas over the past year.

Topic I: Survey and (Sample) Data Collection

In response to an immediate need for a baseline of both bee production and health, several survey and data collection efforts, supported both by ARS and NIFA, have been underway. Together, the different survey efforts have better defined CCD symptoms. Previous studies showed that symptoms included a rapid loss of adult honey bees, excess immature bees present in the combs, and the queen still present. Additional findings indicate an absence of damaging levels of the gut parasite Nosema or parasitic varroa mites at the time of collapse. Data on overall honey bee losses for 2010 indicate an estimated 34 percent loss, which is statistically similar to losses reported in 2007, 2008, and 2009. Survey work will continue to assess the status of honey bee health and further refine CCD symptomology.

The Animal and Plant Health Inspection Service (APHIS) has completed a pilot national survey, which will be expanded this coming year to include sampling in 13 states to look for exotic pests and diseases in honey bee colonies.

The National Agricultural Statistics Service (NASS) has discussed methods and procedures to incorporate colony loss questions into the current NASS national honey production survey. The Environmental Protection Agency (EPA) recently hosted a USDA-led workshop to discuss the development of new toxicity testing protocols to determine the potential effects of pesticides on bees. EPA and USDA officials are members of a steering committee composed of international representatives from government, industry, academia, and non-governmental organizations who are organizing a more comprehensive workshop to be convened at the global Society of Environmental Toxicology and Chemistry (SETAC) Pellston[1] workshop in the coming year. The global Pellston is intended to develop an improved approach for characterizing risks of pesticides to bees and to identify the data needed to inform the risk assessment process. The SETAC conference will also examine the adequacy of tests performed with *Apis mellifera* to serve as a model to predict effects to other "non-Apis" pollinators.

Topic II: Analysis of Existing Samples

Following up on efforts from previous years, researchers supported both by NIFA and ARS continue to analyze bee samples for pesticide residues and pathogen loads to determine possible linkages to bee declines. Studies continue to demonstrate very high levels of pathogens in CCD-affected samples and lower pathogen levels in non-affected samples, consistent with the empirical observation that healthy honey bee colonies normally fend off pathogens. These observations have led to the hypothesis that bee declines are resulting from immune suppression. A large survey of healthy and CCD-affected colonies also revealed elevated levels of pesticides in wax and pollen, but the amounts of pesticides were similar in both failing and healthy colonies.

Supporting the observation that CCD is caused by a combination of factors is a series of studies within NIFA's Coordinated Agricultural Project (CAP). CAP scientists reported several important findings in the past year. In addition to the demonstration that pesticide application rates are highly variable regionally—again pointing away from one specific cause of bee decline—CAP scientists have identified interactive effects among pesticides, the parasite Nosema, and viruses—including the Israeli Acute Paralysis Virus (IAPV) and deformed wing virus, all known stressors to honey bees. Another multi-year, complex study within the CAP project is taking place across multiple states to examine additional interactions among pesticides, pests, and parasites. These studies are just beginning and are yielding working hypotheses based on preliminary, mostly unpublished data. For example, one study indicates potential interactive effects between varroa mites and small hive beetles on colony losses.

CAP-funded studies also identified sub-lethal effects of neonicotinoids[2] and fungicides on bees. It is hypothesized that these pesticides impair the bee's immune system, which leaves the bee more susceptible to three important bee viruses. Future experiments are needed to test these hypotheses and conclusively identify mechanisms of immune response, as well as how these interactions might affect mortality and colony health.

Separate research has independently demonstrated synergistic effects between various pesticides and pathogens. A study by Pennsylvania State University scientists, funded by the National Honey Board, identified potential interactions among numerous pests, pathogens, and pesticides. Scientists also found that one fungicide (chlorothalonil) was linked with "entombing" behavior in bees (a defensive behavior associated with poor health) and had particularly detrimental effects. Studies are ongoing, and as more light is shed on the latent effects of these sub-lethal and synergistic factors, strategies to mitigate these effects may evolve along with the development of management strategies for beekeepers to protect their bees. CAP-funded scientists from Pennsylvania State University recently conducted preliminary studies that indicate IAPV has an East Coast and a West Coast variant. IAPV was believed to be linked with CCD in 2007. The west coast IAPV variant appears to be more closely related to the predominant strain in Australia. Currently, studies are underway to compare the virulence of the East Coast and the West Coast strains of IAPV, and to better understand any association with other factors linked to CCD.

Continuing to capitalize on the honey bee genome, and with funding from NIFA, University of Illinois scientists have identified potential molecular markers for genes that play a role in CCD (Proceedings of the National Academy of Sciences, 2009). Although application of these results is a long way off, these markers will be highly useful for breeding bees with enhanced tolerance to pesticides and/or pathogens associated with CCD and declining bee health in general.

Improved diagnostic assays are ongoing with funding from both ARS and the NIFA-funded CAP to analyze CCD samples within the context of sample analysis and in Topic III, Hypothesis-Driven Research, described below.

Topic III: Research to Identify Factors affecting Honey Bee Health, Including Attempts to Recreate CCD Symptomology

Research efforts jointly supported by ARS and NIFA continue to investigate factors that may play a role in causing CCD, either alone and/or in combination. Factors include diseases (parasites and pathogens), pesticides, poor nutrition, beekeeping practices, and to a lesser extent, other pests such as the small hive beetle.

Certain pesticides are suspected as a contributing factor to CCD. Inclusion of pesticides as a contributing factor to CCD is predicated on the relative paucity of detoxification enzymes (which help insects fend off toxins in the environment) found in CCD samples. It is believed that bees have a relatively weak immune system as a result, and that this may be an evolutionary consequence of feeding on nectar and pollen, which are plant materials lower in toxin levels than foliage. However, it has been noted in the literature that nectar and pollen in certain plant species harbor secondary plant compounds that are toxic to bees. And recently, scientists[3] reported the occurrence of pesticides in pollen. Further studies are needed to determine whether these residues are affecting honey bees either sub-lethally or acutely, and whether any sub-lethal effects may impact the immune response of bees.

Over the past year, given the knowledge of high pathogen loads in diseased bee colonies, ARS- and NIFA-supported researchers have increasingly focused on possible interactive effects between pesticides and pathogens, including Nosema and bee viruses, as noted in the analysis studies under Topic II. Two recent studies have demonstrated increased levels of the gut parasite Nosema following exposure to sub-lethal levels of imidacloprid. As noted previously, a survey of healthy and CCD-affected colonies revealed an elevated level of pesticides in wax and pollen, but the amounts were similar between failing and healthy colonies.

In addition, noteworthy bioassay studies, conducted by NIFA-funded scientists at the University of Illinois at Urbana-Champaign and University of Nebraska at Lincoln, revealed synergistic effects of different chemicals on honey bee health. The studies revealed that these compounds (such as the pyrethroid pesticide fluvalinate, the organophosphate pesticide coumaphos and the fungicide prochloraz, if used in combination) produced an increase in toxicity to bees over what should be expected from use of the products individually. These studies show that the pesticides interfere with the bees' detoxification processes, indicating that these pesticides can be safely used individually, but point to specific pesticide and fungicide combinations that may harm bees and should be avoided. If confirmed in follow-up studies, this information will be highly useful for informing bee health management plans that permit optimal crop pest and disease control with minimal impact on pollinators.

ARS researchers in the Areawide Project on Honey Bee Health have initiated a study to investigate possible interactive effects among pesticide and pathogen load, pests, and diet quality on honey bee health. Results, to become available in spring 2011, will help identify these possible interactive effects.

Researchers have completed sequencing and annotating the genomes of several major bee pests and parasites, including *Nosema ceranae* and *Nosema apis* as well as partial sequencing of the varroa mite genome. With industry support, ARS completed the *Nosema* spp. sequencing effort last year, and work continues with varroa, which is being supported by ARS and NIFA.

Analysis is ongoing to pinpoint important genes and capitalize on specific vulnerabilities that will allow development of better control strategies for these pests and parasites.

Topic IV: Mitigation and Management Measures

Significant progress has been made, and work continues, in the area of mitigating the honey bee crisis. This work is highlighted by two national multi-year projects, the ARS Areawide Project on Honey Bee Health and a NIFA-funded CAP, through which researchers continue to test possible causes of CCD and develop strategies to improve bee management.

Efforts to enhance mite-resistant stocks and breed-resistant bees maintain an important component were conducted in the ARS Areawide Project and NIFA supported research efforts. In particular, research continues to focus on developing germplasm (i.e. bee lines) with the varroa sensitive hygiene (VSH) trait, shown to protect bees against the most significant bee pest, the varroa mite. As these efforts continue, varroa-resistant bee stocks will become increasingly available to beekeepers and will reduce the overall impact of this major known cause of bee decline. Also along these lines, university scientists have several training efforts underway to help bee producers improve breeding and maintenance of healthy bee stocks.

The Areawide Project has also supported several studies informing honey bee diet recommendations for beekeepers. Studies have supported prior findings that supplemental protein feedings can strengthen honey bee colonies, while high-fructose corn syrup may cause colony stress.

Meanwhile, other researchers continue to produce tools and recommendations that help beekeepers limit the impact of common bee pests and pathogens. For instance, ARS researchers have developed chemical attractants to help control the small hive beetle. Other ARS research has shown that ozone—a common means of controlling contamination and rotting of agricultural products—can also be used to kill bee pests and pathogens.

Research is ongoing regarding the development of non-Apis pollinators intended to provide alternatives to honey bee pollination. It is important to note that ARS- and NIFA-funded scientists have made significant progress toward developing the blue orchard bee (*Osmia lignaria*) for almond pollination ("How to Manage the Blue Orchard Bee," Bosch and Kemp, USDA-NIFA Sustainable Agriculture Research and Education (SARE) and ARS, 2001). Development of the blue orchard bee may potentially reduce pressure on honey bee colonies which are the primary pollinators for almond. With funding from SARE, the Minnesota Department of Entomology, and the Xerces Society for Invertebrate Conservation, bee extension specialists and conservationists published a book in 2010 entitled "Managing Alternative Pollinators." This book provides the latest information on the importance of native bees to pollination of agricultural crops.

Pollinator studies have also focused on exploring bumble bees (*Bombus* spp.), seeking to determine the relationship between bumble bee and honey bee pests. Studies have revealed that bumble bees are affected by IAPV and other viruses and parasites that affect the honey bee. These findings may have potential implications for the origin of these factors in bee decline and may assist in understanding and ultimately controlling bee parasites.

To meet the need for an improved regulatory framework for honey bee imports/exports, APHIS and ARS have developed guidelines to coordinate regulation and importation of bees between Mexico, Canada, and the United States, seeking to avoid introducing new bee pathogens and parasites during export.

Also supporting this objective, APHIS has carried out a pilot national pest survey in three states (Hawaii, Florida, and California) and will expand it this coming year to 13 states. This survey will provide increased knowledge of pest and pathogen impacts on bee health and will determine if exotic new pests are present in the United States. Coupled with international collaboration (OIE (World Organization for Animal Health) and COLOSS (Colony Loss Prevention Working Group of the EU)) to identify similar patterns in bee health worldwide, this survey will both provide information that could inform actions to improve pollinator survival and allow the United States to better meet World Trade Organization (WTO) standards with regard to international trade.

To transfer these and other technologies to the field, a number of websites have been created, including the Bee Health Community of Practice eXtension website, which will contain Best Management Practices for beekeepers, along with several scientific publications to relay findings quickly and easily to beekeepers and other users (http://www.extension.org /bee%20health). These resources will facilitate dialogue and enhance information sharing as new discoveries are made regarding bee health and CCD.

Increasingly, CCD results have found a presence on a variety of social networking and public sites, from Facebook to Twitter and YouTube. These newer venues are an innovative mechanism to educate the public and raise awareness of the plight of pollinators.

The Appendix that follows provides specific results and findings from the past year of research, listed within the framework of the Action Plan.

APPENDIX: SPECIFIC ACCOMPLISHMENTS BY ACTION PLAN COMPONENT

Topic I: Survey and Data Collection

Goal 1: Determine the Extent of CCD in the United States

1. Refine CCD symptomology to determine what CCD is and what it is not.

ACCOMPLISHMENT

Definition of CCD. A recent publication (vanEngelsdorp et al. 2009, PLoS ONE, vol. 4, issue 8) described in detail the field symptoms used to define colonies as suffering from CCD.

Symptoms include (1) the apparent rapid loss of adult worker bees from affected colonies as evidenced by weak or dead colonies with excess brood populations relative to adult bee populations; (2) the noticeable lack of dead worker bees both within and surrounding the hive; and (3) the delayed invasion of hive pests (e.g., small hive beetles and wax moths) and kleptoparasitism (stealing food) from neighboring honey bee colonies. A Descriptive Study gives a detailed account of how to diagnose CCD in the field and the use of laboratory tests for varroa mites and nosema that should also be excluded as causative if these parasites are present at low levels. This paper lays the groundwork for comparative studies as it clearly defines the parameters of CCD and thus helps refine the symptoms associated with this disorder and helps to guide hypothesis-driven research designed to determine causation.

FUNDING
National Honey Board
ARS Areawide Program
Pennsylvania Department of Agriculture
Pennsylvania State Hatch Funds
North Carolina Agriculture Foundation
NIFA

PROJECT CONTACTS
Dennis vanEngelsdorp (Dennis.Vanengelsdorp@gmail.com)
Jeff Pettis (Jeff.Pettis@ars.usda.gov)

2. Develop and conduct an expanded, systematic, nationwide, epidemiological survey, based on existing models.

ACCOMPLISHMENT

Colony Losses Remain at High Levels, Similar to Previous Two Beekeeping Seasons. ARS scientists and the Apiary Inspectors of America surveyed U.S. honey bee colonies and estimated total losses, from a variety of causes, to be more than 29 percent from the fall of 2008 through winter of 2009, representing absolute colony losses ranging from 580,000 to 770,000 hives. The 2009-2010 data indicate losses of approximately 34 percent. These data indicate a statistically similar loss rate to the 31 percent and 36 percent losses recorded during the 2006-2007 and 2007-2008 seasons, respectively. The 2008-2009 colony loss survey appeared in a special issue of the Journal of Apicultural Research (January 2010), and a preliminary report of the 2009-2010 study was released in May 2010.

FUNDING
ARS
AIA
Pennsylvania Department of Agriculture

PROJECT CONTACTS
Dennis vanEngelsdorp (Dennis.Vanengelsdorp@gmail.com)
Jeff Pettis (Jeff.Pettis@ars.usda.gov)

Goal 2: Determine Current Status of Honey Bee Colony Production and Health

1. Adapt the annual NASS honey production survey to include information on pollination services and colony losses.

There are currently no resources for this activity at NASS.

2. Develop a long-term annual APHIS survey on the overall health status of U.S. honey bees.

ACCOMPLISHMENT

Workshop Held to Examine the Potential Effects of Pesticides on Honey Bees. APHIS hosted a USDA-funded workshop at EPA with ARS, California Department of Pesticide Regulation, Canadian Pesticide Management Regulatory Agency, and the French Agency for Food Safety, and university representatives to discuss laboratory, semi-field, and field toxicity testing protocols for honey bees and other non-target invertebrate pollinators. With the knowledge that pesticides could potentially have significant effects on bees, EPA recognizes the need for more comprehensive testing of some pesticides and possible revision of past standards. The proceedings of this workshop will be made available in the coming year.

FUNDING
ARS

PROJECT CONTACTS
Robyn Rose (Robyn.I.Rose@aphis.usda.gov)
Jeff Pettis (Jeff.Pettis@ars.usda.gov)
Tom Steeger (Steeger.Thomas@epamail.epa.gov)
Tom Moriarty (Moriarty.Thomas@epamail.epa.gov)

Topic II: Analysis of Existing Samples

Goal 1: Identify and Characterize Pathogens Associated with CCD

1. Analyze samples using:

- High-throughput sequencing for pathogen detection in individual colonies;
- Microarray analysis and quantitative gene expression studies to determine stressor or pathogen effects on bee gene expression; and
- Integrated Virus Detection System (IVDS) for identifying pathogens by particle size.

ACCOMPLISHMENTS

High Virus Prevalence in Local and Migratory Honey Bees Demonstrated. University of Massachusetts researchers analyzed local and migratory honey bee hives from eastern Massachusetts for seven viruses and identified three with a significant presence in the samples. Over 98 percent of the bees were infected with at least one virus; of the seven viruses examined, black queen cell virus (BQCV), deformed wing virus (DWV), and sacbrood virus (SBV), in that order, had a high presence. These results will help researchers pinpoint the factors most likely to play a role in bee decline and will help target development of pathogen control strategies.

Methods Developed to Detect Virus Replication in Honey Bees. The assessment of virus effects on bee colony health requires not only the detection of viruses in bees and bee colonies, but also the indication of virus replication in the host. Scientists at the University of Massachusetts have developed a simple RT-PCR-based assay to detect the most active stage of virus replication— production of a material known as "negative strand, replicative intermediate"—and determined that 50-80 percent of the virus-positive bees had these intermediates. These results indicate that actual virus infection rates were much lower than previously understood.

FUNDING
NIFA
PROJECT CONTACT
John P. Burand (jburand@microbio.umass.edu)

Interactions of Israeli Acute Paralysis Virus (IAPV) and Deformed Wing Virus (DWV) on Honey Bee Health Examined. In a Cooperative Agricultural Project between ARS and the University of Illinois, scientists examined the interactions among IAPV, DWV, pesticides, and Nosema. Results showed that IAPV-infected bees exhibited 90-100 percent early (3-5 day post-infection) mortality rates and paralytic-like seizures before death. Behavioral changes included increased brood feedings and release of Nasanov pheromones, which draw bees back to the hive and are associated with worker recruitment to the hive or food and water sources. These results implicate the virus as being associated with both mortality and behavioral changes, which are possible links to CCD. The behavioral changes, in particular, deserve further investigation since CCD might be a consequence of such changes.

Studies also showed that bees exposed to neonicitinoid insecticides and fungicides at sub-lethal levels had impaired bee immune systems and increased levels of the viruses noted above. Since CCD might be a result of such interactions and damaged bee immunity, further studies are planned to examine immune responses and to determine the mechanisms underlying these interactions.

Research is also characterizing levels of *Nosema ceranae* and *Nosema apis* to determine possible interactions and synergies with the viruses above.

FUNDING
NIFA ARS
PROJECT CONTACTS
Lee Solter (lsolter@illinois.edu)
Judy Chen (Judy.Chen@ars.usda.gov)
Jay Evans (Jay.Evans@ars.usda.gov)

High Pathogen Loads Identified in Diseased Bee Populations. Researchers compared 61 variables (e.g., adult bee physiology, pathogen loads, and pesticide levels) between healthy and diseased bee populations. They found no single CCD causal agent, but rather that CCD colonies had higher total pathogen loads and were infected with a greater number of pathogens than non-CCD colonies, suggesting that bee immune systems had been compromised (vanEngelsdorp et al., PLoS One, 2009). This work will pave the way for future studies to identify specific mechanisms leading to collapse and to reconstitute CCD in laboratory experiments, as well as strategies to mitigate pesticide and pathogen buildup and otherwise protect honey bee health.

FUNDING
National Honey Board
ARS Areawide Program
Pennsylvania Department of Agriculture
Pennsylvania State Hatch Funds
North Carolina Agriculture Foundation
NIFA-AFRI

PROJECT CONTACTS
Chris Mullin, Penn State University (CAMullin@psu.edu)
James L. Frazier, Penn State University (jff2@psu.edu)

Diagnostic Tools Developed for High-Throughput Detection and Monitoring of Honey Bee Diseases. ARS and Pennsylvania State University scientists developed a new diagnostic tool for detection of Nosema infection in field honey bee samples. This new highly specific, sensitive test, based on an antigen capture assay that detects Nosema spore wall protein, is capable of detecting Nosema antibodies. Researchers are pursuing leads to transfer this technology to private industry.

FUNDING
ARS
NIFA

PROJECT CONTACT
Katherine Aronstein (kate.aronstein@ars.usda.gov)

Differential Susceptibility of Bees to IAPV. Honey bee workers of different ages were found to be differentially susceptible to IAPV infection. Newly emerged adult bees were highly susceptible to infection by IAPV with median mortality occurring at 46 h post emergence. The susceptibility declines as bees age, with 5- to 6-day-old bees exhibiting 40 percent mortality at 80 h. Bees parasitized as pupae (with normal wings) by varroa are approximately twice as susceptible to IAPV infections as non-parasitized bees.

FUNDING
USDA-NIFA

PROJECT CONTACT
Diana Cox-Foster (dxc12@psu.edu)

2. Isolate, purify, and quantify microbes associated with CCD.

ONGOING RESEARCH

In Vitro Production of Bee Viruses. Research into the molecular biology, pathology, replication, and transmission of viruses in honey bees has been greatly hindered by lack of an in vitro, cell culture system for the propagation of these viruses. University of Massachusetts researchers are currently developing methods using insect cell lines to produce RNA viruses outside of the bee. At present, scientists have been able to produce double-stranded DNA molecules of deformed wing virus RNA genome by making cDNAs and are developing methods to express viral gene products. This will be useful for determining the role of viruses in CCD.

FUNDING
NIFA

PROJECT CONTACT
John P. Burand (jburand@microbio.umass.edu)

ARS-Ft. Pierce is producing IAPV and other viruses in bee larvae and pupae through microinjection of purified viruses. Progress has also been made in producing bee hemolymph cell cultures, with cells remaining viable for several months. Efforts to culture bee viruses in the cells have begun. If successful, this will provide a fast and readily available system for bee virus production.

FUNDING
ARS

PROJECT CONTACT
Wayne Hunter (Wayne.Hunter@ars.usda.gov)

Goal 2: Determine Pests Associated with CCD and Quantify Pest Levels Associated with the Disorder

Use standard sampling methods to analyze samples for tracheal and varroa mites and Nosema species.

ONGOING RESEARCH

Interactions between Pests, Pathogens, and Pesticides Being Investigated. With funding from the NIFA Coordinated Agricultural Project, scientists are examining the potential interactive effects of pests, pathogens, and pesticides in seven states across the United States. Researchers have been correlating infestation of varroa mite and small hive beetle with colony losses, seeking to determine patterns. To date, research confirms the detrimental effect of pest infestation on colony health. This long and complex study is critically necessary to evaluate the effects of interactions and synergies and bee health, which are little understood but suspected to play a role in CCD.

FUNDING
NIFA

PROJECT CONTACTS
Frank Drummond (frank.drummond@umit.maine.edu)
Nancy Ostiguy (nxo3@psu.edu)
Katherine Aronstein (kate.aronstein@ars.usda.gov)
Steve Sheppard (shepp@wsu.edu)
Kirk Visscher (visscher@ucr.edu)
Marla Spivak (spiva001@umn.edu)
Jamie Ellis (jdellis@ufl.edu)

Goal 3: Identify Pesticides or Environmental Contaminants Associated with CCD

1. Examine wax, pollen, honey, and adult bee samples for pesticides and environmental contaminants.

and

2. Determine whether interactions between pesticides applied inside bee hives and pesticides applied to crops contribute to CCD.

ACCOMPLISHMENTS

Link Between Queen Status, Brood Pattern, Varroa/Nosema, Pesticides, and Colony Health Studied. ARS and Pennsylvania State University (PSU) researchers are completing an epidemiological analysis of the relative risk of various conditions in colonies in three migratory operations. In a study of pesticide residues in bee colonies across 23 states, PSU researchers found numerous pesticides and metabolites in the hive samples studied, increasing the opportunity for toxic interactions between the chemicals. Initial analysis discovered a relatively new condition in bee colonies—termed "entombed pollen"—a defensive behavior correlated with poor health and increased mortality rates (Journal of Invertebrate Pathology, 2009). This condition is thought to be the result of the fungicide chlorothalonil in colony pollen stores. Data analysis continues, with queen failure and diseased brood being critical predictors of colony mortality.

FUNDING
National Honey Board
Project Apis mellifera
Foundation for the Preservation of Bees
Pennsylvania Department of Agriculture
NIFA

PROJECT CONTACTS
Jeff Pettis (Jeff.Pettis@ars.usda.gov)
Jay Evans (Jay.Evans@ars.usda.gov)
Dennis vanEngelsdorp (Dennis.Vanengelsdorp@gmail.com)
Chris Mullin, Penn State University (CAMullin@psu.edu)
James L. Frazier, Penn State University (jff2@psu.edu)

Pesticide Variability Analyzed. Scientists supported by NIFA's Coordinated Agricultural Project analyzed pesticide residues in bee hives and identified a variety of different chemicals and their metabolites in hives. Results indicate that exposure is highly variable depending on sampling location and that further research is necessary to understand the spatial and temporal variability.

FUNDING
NIFA

PROJECT CONTACTS
Frank Drummond (frank.drummond@umit.maine.edu)
Nancy Ostiguy (nxo3@psu.edu)
Katherine Aronstein (kate.aronstein@ars.usda.gov)
Steve Sheppard (shepp@wsu.edu) Kirk Visscher (visscher@ucr.edu)

Insecticide Novaluron Found to Have Limited Effects on Bee Eggs. In studies on female alfalfa leaf-cutting bees, ARS scientists evaluated whether the insecticide novaluron could affect the health of bee eggs and larvae produced by exposed mother bees. Although bees fed high levels of the pesticide experienced significantly greater larval losses than unexposed bees, in follow-up studies using typical field treatment levels of the pesticide, mortality rates were similar for unexposed and exposed bees.

FUNDING
ARS
NIFA

PROJECT CONTACTS
Theresa Pitts-Singer (Theresa.Pitts_Singer@ars.usda.gov)
James Barbour (jbarbour@uidaho.edu)
Erin Hodgson (erin@biology.usu.edu)

Rimon Affects Bee Egg Viability. Novaluron (Rimon) is a relatively new insecticide that is an insect growth regulator, labeled for lygus control in alfalfa fields during bloom. Some alfalfa leaf-cutting beekeepers have complained about poor bee returns in fields where novaluron was used, and laboratory trials have shown this pesticide to be toxic to larvae and eggs. Field experiments were conducted in 2009, in a collaboration between the University of Idaho and ARS, to evaluate the effect of novaluron (Rimon 0.83 EC) on alfalfa leaf-cutting bees. The mean number of bee cells in plots treated with novaluron and plots treated with another insecticide typically used by growers did not differ, but both the total number and the percent of pollen balls (an early stage mortality, or eggless cells) was higher in plots treated with novaluron, with about 20 percent more pollen balls in the novaluron treatment. The percentage of live larvae was also lower in novaluron-treated fields. Dissection of pollen balls revealed that plots treated with novaluron had a significantly lower percentage of live eggs. Novaluron did not appear to affect adult bee survival or nesting activity, but had a negative impact on egg viability and the number of live larvae at the end of the season.

FUNDING
ARS
NIFA
Chemtura Inc.

PROJECT CONTACTS
Jim Barbour (jbarbour@uidho.edu)
Theresa Pitts-Singer (Theresa.Pitts_Singer@ars.usda.gov)

Effects of Pesticides on Alfalfa Pollinators Studied. In large-scale field trials carried out by ARS scientists cooperating with Washington State University, scientists evaluated the effects of several insecticides on pollinators.

FUNDING
Washington State Alfalfa Seed Commission
Washington State Commission on Pesticide Registration
ARS

PROJECT CONTACTS
Douglas Walsh (dwalsh@wsu.edu)
Rosalind James (Rosalind.James@ars.usda.gov)

ONGOING RESEARCH

Stationary Apiary Health Parameters Comparisons. In the Coordinated Agricultural Project (CAP), scientists assessed the presence and quantity of tracheal and varroa mites, viruses, Nosema, and pesticides in wax, as well as quantity of pollen, queen health parameters, and brood and adult bee populations, evaluating possible linkages between pest/pathogen/pesticide presence and colony survivorship. Data show high variability among virus presence, and evaluations continue to identify possible linkages. There was also a high variability in the pesticide load in the pollen brought back to the hives.

Pesticide Levels in Honey Examined. CAP scientists are evaluating the type and level of pesticides in honey extracted from bee hives and in honey that overwinters with the bees. This overwintered honey is the source of carbohydrates for the bees during winter and pesticides present in the honey will be ingested by the bees. The miticides coumaphos and fluvalinate were observed more frequently and at higher concentrations than any other pesticide, and results indicate that coumaphos may have migrated from brood chamber use.

FUNDING
NIFA

PROJECT CONTACTS
Nancy Ostiguy (nxo3@psu.edu)
Brian Eitzer (Brian.Eitzer@ct.gov)

Goal 4: Develop Analytical Tools to Assess Bee Health

Develop the use of molecular markers to determine the physiological status of bees and as indicators of bee health.

ACCOMPLISHMENT

Honey Bee Detoxification Genes Identified. Certain chemicals commonly found in plants may be toxic to bees, and University of Illinois researchers have found that genes in the honey bee genome may have a role in processing chemicals produced by the plant and encountered by honey bees when they feed on nectar. Recent studies revealed that several genes inside bees were activated when fed extracts of honey, pollen, and propolis. Findings indicate the possibility that genetic differences among honey bees could explain differences in their ability to detoxify compounds in nectars with high flavonoid content. This information could help beekeepers breed strains of honey bees that tolerate plants with toxic compounds in nectar and pollen.

Genes Associated With Pesticide Metabolism Identified. Using the honey bee genome sequence, University of Illinois researchers have identified genes that facilitate metabolism of pesticides. Genetic analysis showed that certain enzymes metabolized the varroa mite-killing product Apistan (fluvalinate). Work is ongoing to characterize the metabolites, and molecular modeling will help identify key enzymes involved in breaking down pesticides. This work may potentially explain why certain pesticides are relatively benign in the context of acute toxicity to bees. This research is consistent with prior findings when Apistan (fluvalinate) is not combined with other pesticides (see Goal 4: Elucidate the synergistic and sub-lethal effects of in-hive miticides on colony health), and could be used to assist in breeding efforts to develop more bees that may tolerate or resist pesticide exposure.

FUNDING
NIFA

PROJECT CONTACT
May R. Berenbaum (maybe@illinois.edu)

Topic III: Research to Identify Factors Affecting Honey Bee Health, Including Attempts to Re-create CCD Symptomology

Goal 1: Confirm or Eliminate Potential Environmental Stressors as Contributing Causes of CCD

1. Test effects (lethal and sub-lethal) of neonicotinoids and other pesticides used for crop protection.

ACCOMPLISHMENTS

Synergistic Effects of Pesticide Exposure on Pathogen Growth and Bee Health Discovered. There has been significant concern about possible interactive effects of various pesticides and pathogens on bees. When bees were exposed to the gut parasite Nosema and insecticide imidacloprid, researchers found evidence of an interactive effect on honey bee health. Worker bees exhibited up to a fourfold increase in Nosema levels when they originated from colonies that had been fed imidacloprid, indicating a subtle sub-lethal interaction between pesticides and pathogens. This research has been submitted for publication, and similar findings have recently been published from a study in France in Environmental Microbiology 2009.

FUNDING
ARS
NAPPC

PROJECT CONTACTS
Jeff Pettis (Jeff.Pettis@ars.usda.gov)
Galen Dively (Galen@umd.edu)
Dennis vanEngelsdorp (Dennis.Vanengelsdorp@gmail.com)

Interaction between Pesticides and Bee Viruses Found. Pennsylvania State University researchers found that consumption of sub-lethal levels of certain pesticides elevates viral titers in infected bees and appears to alter the immune responses of the bees. In cage studies, nurse or house bees were fed sugar solutions containing pesticides (myclobutanil, acetamiprid, fluvalinate or chlorothalonil) at dosages previously determined to be sub-lethal. These dosages approximated lower concentrations found in incoming pollen. The titers of several picorna-like viruses (deformed wing virus, black queen cell virus, and IAPV) were found to be elevated by one or several of the pesticides. The expression of several immune-related genes was altered by the pesticide exposure as compared to untreated bees that were also immune challenged. The identity of the viruses was confirmed via genome sequencing.

FUNDING
Pennsylvania Department of Agriculture.

PROJECT CONTACT:

Diana Cox-Foster (dxc12@psu.edu)

Movement of Imidacloprid in Trees Explored. APHIS has used soil and trunk injection of imidacloprid to treat hardwood trees in controlling the invasive Asian longhorned beetle. There is the potential for movement of the pesticide into pollen and nectar of treated trees. APHIS is collaborating with ARS to monitor the movement of imidacloprid into pollen and nectar and to evaluate the potential effects on honey bees and other non-target species over a 3-year period.

FUNDING

APHIS

ARS

PROJECT CONTACTS

Jeff Pettis (Jeff.Pettis@ars.usda.gov)

Robyn Rose (Robyn.I.Rose@aphis.usda.gov)

2. Test the effects of current miticides used in hives on worker bee longevity and colony health.

ACCOMPLISHMENTS

Reduced Longevity of Adult Worker Bees Related to Coumaphos Exposure in Wax During Larval and Pupal Development. There was a significant reduction in adult bee longevity following exposure to 100 ppb of coumaphos in wax during the larval and pupal stages in worker honey bees. A 4-day reduction in summer bee lifespan was observed equaling 16 percent of the total lifespan of summer bees. Reduced adult longevity could impact honey production and or overwintering ability. This study has been accepted for publication in Apidologie.

FUNDING

ARS

Pennsylvania Department of Agriculture.

PROJECT CONTACTS

Jeff Pettis (Jeff.Pettis@ars.usda.gov)

Dennis vanEngelsdorp (Dennis.Vanengelsdorp@gmail.com)

3. Test the effects of antibiotics (especially new ones such as Tylosin) on the increase in pathogens (e.g., *Nosema ceranae*) and the overall viability of bees over winter.

No accomplishments to report.

4. Test effects of supplemental protein and carbohydrate (e.g., high-fructose corn syrup (HFCS)) feedings on bee health.

See Topic IV, Goal 3, Objective 2.

ONGOING RESEARCH

Plant Resin Identified to Improve Bee Resistance to Fungal Toxins. Honey bees and their resource-rich nests are hosts to a wide range of fungi, including species that produce toxic substances called mycotoxins, which workers detoxify using gut enzymes. In bioassay experiments, bees that consumed extracts of propolis, a complex mixture of plant resins collected by bees and used as a general caulking material in the hive, had an enhanced capacity to detoxify mycotoxins. These results suggest that the enzymes involved in mycotoxin detoxification are enhanced by ingestion of propolis, which may serve a hitherto unrecognized role in honey bee health.

PROJECT CONTACT

May R. Berenbaum (maybe@illinois.edu)

5. Test effects of availability and quality of natural food sources on bee health as affected by climatic factors (e.g., drought).

6. Test effects of management practices (e.g., nutrition, migratory stresses) on bee health. These objectives are being done as part of the Areawide project. See Part IV-6.

Goal 2: Confirm or Eliminate Potential Pathogens as Contributing Causes of CCD

1. Test pathogenicity of the following CCD-associated microbes against honey bees and non-Apis bees:

- Viruses
- Fungi (chalkbrood; stonebrood)
- Microsporidia (Nosema)
- Bacteria
- Trypanosomes and other microbes

ACCOMPLISHMENTS

Viral Diseases and Varroa Mites Linked to Colony Losses. Since 2008, ARS scientists in Weslaco have collaborated with university scientists to determine key factors (e.g., parasites, diseases, and pesticides) contributing to honey bee colony losses in apiaries across the United States. First-year data from this Coordinated Agricultural Project (CAP) indicate that varroa mite infestation and viral diseases contribute strongly to colony losses.

FUNDING

ARS

NIFA

PROJECT CONTACT

Katherine Aronstein (Kate.Aronstein@ars.usda.gov)

2. Compare genes expressed in response to specific pathogens or pesticides with those expressed in bees from CCD colonies.

Varroa Mite and Nosema Species Being Sequenced. ARS scientists in Beltsville have finished sequencing, annotating, and publishing the *Nosema ceranae* genome and have sequenced *Nosema apis*, with annotation and publication expected sometime in 2010. Also, with industry support, ARS initiated a genome project on Varroa destructor, the primary honey bee parasite, as part of a worldwide consortium. Further analysis of these results will allow researchers to identify genes implicated in virulence, viral disease, and chemical resistance and to develop new strategies to control their impact on bees.

FUNDING
ARS
NIFA

PROJECT CONTACTS
Jay Evans (Jay.Evans@ars.usda.gov)
Judy Chen (Judy.Chen@ars.usda.gov)

Goal 3: Confirm or Eliminate Pests as Contributing Causes of CCD

1.Test the effects of varroa mites on bee health and robustness, particularly overwintering effects and association with CCD in early spring.

No results to report at this time.

2. Determine the importance of varroa as a vector of viruses associated with CCD or as a general immuno-suppressive agent on the colony itself.

No results to report at this time.

Goal 4: Determine What Factors (or Interactions Between Factors) Are Most Important in Their Contribution to CCD. This Includes Environmental Factors (E.G., Temperature, Humidity, and Chemical Exposure), Pathogens and Parasites, and Bee Genetics and Breeding

ONGOING RESEARCH

Interactive Effects of Pollen Quality, Pesticides, Pests, and Pathogens Being Studied. As part of the ARS Areawide Project on Honey Bee Health, researchers from ARS-Beltsville, U.S. Geological Survey-North Dakota, and the University of Minnesota are investigating the interactive effects of pollen quality and availability, pesticide load, and relevant pests and pathogens on honey bee health. Researchers are comparing the performance of honey bee colonies in selected sites in North Dakota during the summer with their ability to overwinter and be available for almond pollination in the early spring. Results will help identify possible interactions between bee stressors and colony health.

FUNDING
ARS
USGS
PROJECT CONTACTS

Jeff Pettis (Jeff.Pettis@ars.usda.gov)
Marla Spivak (spiva001@umn.edu)
Chip Euliss (ceuliss@usgs.gov)

Interactive and Sub-lethal Effects of In-hive Miticides Evaluated. Scientists at the University of Nebraska-Lincoln have been studying the synergistic effects of various varroacides and fungicides on honey bee health, testing for lethal and sub-lethal effects. Research indicated that combinations of varroacides that are detoxified by the cytochrome P450 system (e.g., the miticides tau-fluvalinate, coumaphos, and fenproximate) tend to be significantly more toxic when applied in combination. A similar interaction occurred when bees were exposed to a cytochrome P450-inhibiting fungicide and a varroacide that was metabolized by P450s concurrently. Normally these varroacides are well tolerated by honey bees, but pre-treatment with a P450-inhibiting fungicide greatly increases mortality in bees by a synergistic ratio of nearly 2000X in the most extreme case. While this interaction was found using fungicide doses substantially higher than would likely occur through consumption of fungicide-contaminated pollen (Mullin et al. 2010, PLoS ONE 5: e9754) the potency of this interaction raises concerns about lethal and sub-lethal effects that may be occurring at field-relevant doses. The results suggest that honey bees should not be treated with more than one P450-detoxified varroacide concurrently or sequentially since varroacides are known to accumulate in beeswax. They also suggest that beekeepers should avoid applying P450-detoxified varroacides when honey bees are placed in orchards or other crop settings where exposure to P450-inhibiting fungicides is likely. These results may be useful for incorporation into a bee health management plan. Additional studies are needed to determine the range of bee exposure to fungicides and varroacides in the field and how these concentrations affect the bees.

FUNDING
NIFA

PROJECT CONTACT
Marion Ellis (mellis3@unl.edu)

Topic IV: Mitigation and Preventative Measures

Goal 1: Develop Best Management Practices for Honey Bees

1. Develop best management practices for migratory beekeeping.

ONGOING RESEARCH

Best Management Practices Being Established for Bee Conservation. In July 2009, NIFA's Coordinated Agricultural Project (CAP) and the USDA-ARS Areawide Project launched the Bee Health CAP website (www.beeccdcap.uga.edu), a university eXtension network designed to serve as a clearinghouse for objective, research-based information. The site provides a platform for communities of experts to share articles and resources for download. A Best Management Practices Guide for honey bee health is in progress. For more information on the eXtension site, see Goal 7 (page A-30).

FUNDING
NIFA
ARS

PROJECT CONTACTS
John Skinner (jskinner@utk.edu)
Keith Delaplane (ksd@uga.edu)

Adoption of Sustainable Bee Management Practices Studied. Researchers are surveying beekeepers to determine how many implement best management practices in their operations. Using a web-based assessment, beekeepers are being evaluated in the winter of 2009-2010 and again in 2012. This will provide critical information about how well scientists and beekeepers are able to improve the needs of the bee industry.

FUNDING
NIFA

PROJECT CONTACTS
Thomas Webster (thomas.webster@kysu.edu)
John Skinner (jskinner@utk.edu)
Keith Delaplane (ksd@uga.edu)
Steve Sheppard (shepp@wsu.edu)
Marion Ellis (mellis3@unl.edu)
Frank Drummond (frank.drummond@umit.maine.edu)

2. Develop best management practices for pest and pathogen control.

ACCOMPLISHMENTS

High Ozone Concentrations Found to Kill Insect Pests, Degrade Pesticides, and Kill Pathogen Spores. High concentrations of ozone have been used for many years to decontaminate water and to control rot in stored agricultural products. ARS carried out experiments to determine the concentrations needed to decontaminate honeycomb containing wax moths, pesticides, and pathogens. Ozone was very effective in killing insect pests of stored comb, primarily the wax moth, and was somewhat effective in removing pesticide residues, but very high concentrations, used in combination with heat and humidity, were needed to eliminate pathogens like foulbrood. These studies will assist in developing effective ozone treatment strategies to reduce possible bee stressors in used honeycomb.

FUNDING
ARS

PROJECT CONTACT
Rosalind James (Rosalind.James@ars.usda.gov)

Chemicals Associated With Mite Control Identified. ARS scientists in Gainesville, Florida, have identified two chemicals produced by larval bees that cause reproductively mature mites to move from adult nurse bees into larval cells. These chemicals may be useful as lures to trap varroa mites and disrupt their ability to find larvae. ARS is in the process of patenting the chemicals, which have significant potential to help manage these pests.

FUNDING
ARS

PROJECT CONTACT
Peter Teal (Peter.Teal@ars.usda.gov)

Trap Developed for Small Hive Beetle. ARS-Gainesville scientists have developed and demonstrated the effectiveness of a trap for the small hive beetle. The trap can be placed at the entrance to bee hives to capture larvae of the small hive beetle, thus breaking the cycle of beetle infestation of hives by removing larvae without the need for pesticides. The trap has been demonstrated at beekeeper industry meetings as an effective new pest management technology.

FUNDING
ARS

PROJECT CONTACT
Peter Teal (Peter.Teal@ars.usda.gov)

Expanded Small Hive Beetle Range Demonstrated. The small hive beetle was believed to be dependent primarily on bee pollen, honey, and brood for nutrition, and control efforts focused primarily on areas near honey bee hives. However, ARS scientists have found that the beetle occurs in woodlands far from managed honey bee colonies and that the bee pest can grow on a wider range of foods than previously thought. These findings indicate that the pest's expanded range may make it much more difficult to control and will aid scientists in improving their control strategies.

FUNDING
ARS

PROJECT CONTACT
Peter Teal (Peter.Teal@ars.usda.gov)

Small Hive Beetle Monitoring and Management Techniques Enhanced. Monitoring the small hive beetle is necessary to manage this pest effectively and mitigate the damage it causes. ARS scientists in Gainesville have found that a trap used to monitor the small hive beetle is most effective when placed in shaded rather than open locations, even if located farther from bee hives. These measures will be used by beekeepers to increase the effectiveness of their monitoring and management efforts.

FUNDING
ARS

PROJECT CONTACT
Peter Teal (Peter.Teal@ars.usda.gov)

Chemical Attractants for Small Hive Beetle Identified. In cooperation with the University of Florida, ARS scientists identified volatile chemicals associated with bee samples that were attractive to the small hive beetle. These chemicals could be useful in developing better management strategies for the pests.

FUNDING
ARS

PROJECT CONTACT
Peter Teal (Peter.Teal@ars.usda.gov)

Non-Chemical Control for Small Hive Beetle in Development. ARS scientists in Weslaco signed a new agreement with Mexican collaborators to develop new non-chemical control methods for the small hive beetle. Researchers will test purified Bt toxins shipped from Mexico for effectiveness in controlling beetle populations in bee larvae.

FUNDING
ARS

PROJECT CONTACT
Katherine Aronstein (Kate.Aronstein@ars.usda.gov)

Chalkbrood Resistance Genes Discovered. ARS scientists in Weslaco have discovered genes involved in honey bee resistance to chalkbrood, a major disease affecting honey bees. This finding may lead to the development of new gene-specific pharmaceuticals to fight this disease.

FUNDING
ARS

PROJECT CONTACT
Katherine Aronstein (Kate.Aronstein@ars.usda.gov)

Major Bee Pathogen Sequenced. Recently, ARS-Weslaco scientists finished sequencing one of the major honey bee pathogens, *Ascosphaera apis*, the causative agent of chalkbrood disease. Research focuses on discovering genes involved in the invasion and pathogenesis of the bee host. This information will lead to the development of new disease control methods.

FUNDING
ARS

PROJECT CONTACT
Katherine Aronstein (Kate.Aronstein@ars.usda.gov)

3. Establish guidelines for floral gardens to maintain stronger honey bees.

Goal 2: Develop Best Management Practices for Non-Apis Bees to Provide Alternative Pollinators for Crops, Gardens, and Natural Areas

1. Develop best management practices for pest and pathogen control in non-Apis bees.

ACCOMPLISHMENTS

New Detection System Developed for Studying Disease Transmission. ARS developed a new detection system to assess the number of chalkbrood spores in bee samples. Using this system, scientists evaluated contamination levels in various bee management systems and studied the ecology of disease transmission in farmers' fields. Results will be used to develop improved strategies for disease management.

Immune Systems of the Alfalfa Leaf Cutting Bee Studied. Honey bee immune systems are known to be weaker and smaller than those in other insects, making them more susceptible to pests and diseases. After developing a library of expressed genes from healthy and chalkbrood-infected alfalfa leaf-cutting bees, ARS scientists identified 116 immune response genes for this solitary bee. This finding will allow scientists to greatly improve knowledge of disease resistance in bees—about which very little is known—in order to develop better disease control methods.

High Concentrations of Ozone Found to Kill Chalkbrood. High concentrations of ozone have been used for many years to decontaminate water, and to control rot in stored agricultural products. ARS conducted experiments to determine the concentrations needed to decontaminate leaf-cutting bee nesting boards. The primary pathogen of this bee—chalkbrood—was eliminated with ozone concentrations obtainable in on-farm fumigation chambers, making this a promising strategy for promoting bee health.

FUNDING
ARS

PROJECT CONTACT
Rosalind James (rosalind.james@ars.usda.gov)

Honey Bees and Israeli Acute Paralysis Virus (IAPV) and Deformed Wing Virus (DWV) Studied. Researchers found that IAPV infects bumble bees and reduces colony lifespan. Bumble bees dying from IAPV infections returned to the colony to die inside the hive, unlike honey bees, which died outside the hive. Bumble bees, however, also exhibited the same paralytic-like seizures as observed in honey bees.

Non-Apis Bee Pathogens Identified and Characterized. University scientists have been studying several parasites associated with bumble bees—*Crithidia bombi* (a protozoan parasite of bumble bees), *Nosema bombi*, and the tracheal mite—to determine their transmission patterns, occurrence and distribution, and origin. Although the protozoan was found in commercial bumble bees, it does not seem to be transmitted from wild bumble bees and managed honey bees in the area of Massachusetts investigated. For the protozoan, sequencing studies indicate great similarity between American bumble bees and European bees. Results from these studies will assist in understanding, and, ultimately, controlling bee parasites.

FUNDING
NIFA

PROJECT CONTACTS
Anne Averill (aaverill@ent.umass.edu)
Lee Solter (lsolter@illinois.edu)
Frank Drummond (frank.drummond@umit.maine.edu)

Lethal and Sub-lethal Effects of Insecticides on Non-Apis Bees Elucidated. Using totally enclosed field cage systems, University of Massachusetts scientists assessed the impact of imidacloprid in flowering low bush blueberry on commercial bumble bees (*Bombus impatiens*). After bloom ended, the bees were released to forage on surrounding plants for the remainder of the spring and summer. Results suggest that imidacloprid reduced brood (immature bees) at the end of bloom but did not seem to affect the survival of adult workers.

FUNDING
NIFA

PROJECT CONTACT

Frank Drummond (frank.drummond@umit.maine.edu)

2. Establish guidelines for maintaining stronger populations of non-Apis bees in agricultural systems, home gardens, and wild lands.

ACCOMPLISHMENTS

Blue Orchard Bees Developed to Supplement Honey Bees for Almond Pollination. ARS scientists developed guidance for the management of blue orchard bees for almond pollination and initiated a demonstrated project with university extension personnel. Guidance on managing blue orchard bees for almonds is now available to promote use of this bee and reduce pressure on honey bee colonies.

FUNDING
ARS
University of California-Davis.

PROJECT CONTACTS
James Cane (Jim.Cane@ars.usda.gov)
Theresa Pitts-Singer (Theresa.Pitts-Singer@ars.usda.gov)

Carolyn Pickel (cxpickel@ucdavis.edu)

Alfalfa Leaf Cutting Bee Improved With Proper Stocking Density. The alfalfa leaf cutting bee is commonly used to pollinate alfalfa for seed production in the United States and Canada, but these bees frequently have problems with disease, parasites, predators, and unexplained mortality of eggs and small larvae. ARS researchers have shown that typical release rates for alfalfa leaf cutting bees in alfalfa pollination are too high, decreasing bee efficiency in pollinating flowers and causing them to produce fewer healthy offspring. When too many bees are placed in fields, they compete for food and nesting sites, which results in poor bee returns. Research provides information on optimal numbers of bees needed for effective pollination.

FUNDING
ARS

PROJECT CONTACT
Theresa Pitts-Singer (Theresa.Pitts-Singer@ars.usda.gov)

Historic Range Maps for Native Bumble Bees Constructed. Currently, at least 5 North American bumble bee species are disappearing, with one species thought to have gone extinct within the last 5 years. Despite the rapid loss of these important pollinators, little is known about the cause of their demise. ARS scientists developed a method for determining the extent and cause of these losses using geographic information systems (GIS) and historic records of bumble bee collections, and created models that can predict the geographic range of the bees at different points of time in history. This new method could be used to help scientists understand the effects of climate change or the accidental introduction of pathogens on other pollinators as well.

FUNDING
ARS

PROJECT CONTACT
James Strange (James.Strange@ars.usda.gov)

Fire Shown to Increase Wild Pollinator Diversity. Controlled burns constitute one method used to manage fuel loads in natural areas, but the effects on pollinators remains unknown. ARS scientists completed a 2-year study in Zion National Park, home to 474 different kinds of bees, comparing bee pollinators in adjacent plots of burned and unburned forest and shrub land. In most cases, bee abundance was significantly greater in the burned areas 2 to 5 years after the burn. Burns thus represent opportunities for increasing pollinator populations during forest regeneration.

FUNDING
ARS

PROJECT CONTACT
Terry Griswold (Terry.Griswold@ars.usda.gov)

Economic Impact of Alternative Pollinators Evaluated. ARS scientists have developed a supply and demand model to evaluate the economic impact of introducing a new pollinator (in addition to honey bees) into orchard systems for almond and cherry production. Due to the differences in these crops, they are differentially affected by increases in production acreage and bee supply, and these results will help determine how alternative pollinators can be used effectively to reduce demand for honey bees.

FUNDING
Utah State University
ARS

PROJECT CONTACTS
Ruby Ward (ruby.ward@usu.edu)
Rosalind James (Rosalind.James@ars.usda.gov)

Goal 3: Maintain Bees with Resistance to Parasites and Pathogens

1. Identify traits associated with resistance to parasites and pathogens.

ACCOMPLISHMENT

USDA Varroa-Resistance Stocks Found Well Suited for Commercial Pollination. In a year-long examination of USDA's varroa-resistant stocks for commercial pollination of almonds, apples, blueberries, and cranberries, ARS scientists tested Russian honey bees and bees containing a varroa-resistant trait (varroa sensitive hygiene, VSH) for size and productivity. At every pollination set, researchers found USDA stocks to be as large and productive as control stocks, making them well suited for commercial pollination.

FUNDING
ARS

PROJECT CONTACTS
Tom Rinderer (Tom.Rinderer@ars.usda.gov)
Bob Danka (Bob.Danka@ars.usda.gov)

ONGOING RESEARCH

Improved Management Strategies for Russian Honey Bees Being Developed. ARS scientists are evaluating management procedures for pollination of Russian honey bees. Scientists are studying how quality and quantity of feed, hive size, and winterizing impact colony health and development. When optimized, these procedures will be integrated into the breeding program for improved bee health.

FUNDING
ARS

PROJECT CONTACTS
Tom Rinderer (Tom.Rinderer@ars.usda.gov)
Bob Danka (Bob.Danka@ars.usda.gov)

Mite-resistant Bees With Varroa Sensitive Hygiene (VSH) Trait Selected For Migratory Crop Pollination. ARS breeders at Baton Rouge are developing a population of bees with the VSH trait, which makes them less susceptible to the varroa mite. Research shows that these bees perform well when used in intensive migratory crop pollination. This selected germplasm is expected to be released for use by commercial beekeepers to improve the resistance of their bees to this troublesome pest.

FUNDING
ARS

PROJECT CONTACTS
Bob Danka (Bob.Danka@ars.usda.gov)
Tom Rinderer (Tom.Rinderer@ars.usda.gov)

See Goal 4, Objective 1: Identifying genes that confer resistance to Varroa and pathogens, and genes that respond to biotic challenges.

FUNDING
NIFA

PROJECT CONTACTS
Greg Hunt (ghunt@purdue.edu)
Marla Spivak (spiva001@umn.edu)
Thomas Webster (thomas.webster@kysu.edu)
Katherine Aronstein (Kate.Aronstein@ars.usda.gov)

2. Introduce resistance traits into bee stocks favored by the industry.

Breeding for Hygienic Behavior. University of Minnesota researchers and extension specialists are evaluating commercial honey bee stocks for the level of hygienic behavior, a trait which is associated with the bee's resistance to varroa mite. A pilot certification program will be initiated for the hygienic trait. A "Tech Transfer" team is forming in California to assist bee breeders with stock selection for hygienic behavior and to reduce chemical use in breeder colonies. It is envisioned that this team will serve as a model for the Nation in producing local, regionally adapted queens that are resistant to mites.

FUNDING
National Honey Board
California Almond Board

PROJECT CONTACT
Marla Spivak (spiva001@umn.edu)

ONGOING WORK

Genetic Techniques Used to Improve Bee Breeding. The University of Tennessee has replaced half its bee stock with varieties resistant to varroa mites (either the Varroa Sensitive Hygiene (VSH) strain or with hybrids raised from VSH queens). Equipment and apiary expansion is underway to conduct bee breeding demonstration workshops in 2011 and 2012. Workshop materials will be obtained or developed for inclusion on www.eXtension.org for use in any bee breeding workshop. These efforts will increase the use of robust bee stocks.

FUNDING
NIFA

PROJECT CONTACTS

Nick Calderone (nwc4@cornell.edu)
Keith Delaplane (ksd@uga.edu)
Greg Hunt (ghunt@purdue.edu)
Steve Sheppard (shepp@wsu.edu)
Marla Spivak (spiva001@umn.edu)
Thomas Webster (thomas.webster@kysu.edu)
Michael Wilson (mwilso14@utk.edu)

Groundwork Established for a Sustainable Market for Genetically Improved Queens. Researchers are working to train beekeepers in stock selection (e.g., VSH bees) and breeding techniques, as well as management and testing practices to reduce impact from pests, pathogens, and chemicals. Specifically, scientists are teaching a 2-day Queen Rearing Short Course every summer at the University of Minnesota to assist bee breeders. This course combines lecture and hands-on field work.

In 2008 and 2009, researchers conducted on-site visits with 17 queen producers in northern California. These operations produce at least half of the commercially available queens in the Nation and thus have major impact on honey bee genetics and improvement nationally. Scientists demonstrated how to test their stock for hygienic behavior, how to test their colonies for Nosema, and provided best management practices, including reducing chemical inputs into their breeding stocks.

Washington State University and University of California at Davis researchers have imported selected stock for two Carniolan and Italian bees for breeding purposes. In the future, the team will bring in a third subspecies favored by beekeepers, the Caucasian.

Over the 2-year study, two things became very evident: 1) bee breeders are very competent and produce high-quality, well-mated queens to supply beekeepers throughout the Nation; 2) breeding for pathogenic resistance is highly complex, while maintaining productive characteristics and pollination efficiency requires professional assistance to help bee breeders improve stock selection, enhance genetic diversity, and perform disease diagnostics.

FUNDING
NIFA

PROJECT CONTACTS
Keith Delaplane (ksd@uga.edu)
Marla Spivak (spiva001@umn.edu)

3. Use genomic technologies and germplasm preservation to maintain quantities of desirable honey bee germplasm.

No accomplishments to report at this time.

4. Transition to mite- and pathogen-resistant honey bee stocks. See Topic IV, Goal 3, Objective 2.

Goal 4: Develop Ways to Manage Mite Resistance to Miticides and Create Alternatives

1. Develop resistance management programs that provide beekeepers with tools for mite management.

ACCOMPLISHMENT

Genetic Analysis Completed and Tools Developed. Scientists are studying candidate genes that influence grooming behavior and confer resistance to varroa mites, allowing us to eventually identify genes that influence resistance and potentially lead to marker-assisted selection to speed breeding for varroa mite-resistant bee colonies. Scientists also developed a new laboratory assay to speed the genetic selection for the Varroa Sensitive Hygiene (VSH) trait, which confers resistance to honey bees.

FUNDING
NIFA
ARS

PROJECT CONTACTS
Greg Hunt (ghunt@purdue.edu)
Marla Spivak (spiva001@umn.edu)
Thomas Webster (thomas.webster@kysu.edu)
Katherine Aronstein (Kate.Aronstein@ars.usda.gov)
Christina Grozinger (cmgrozinger@psu.edu)

2. Develop new methods of managing parasites and pathogens.

ONGOING RESEARCH

Identifying Genes That Confer Resistance to Varroa and Pathogens, and Genes That Respond to Biotic Challenges

See Goal 4, Objective 1 above.
Deliverables for this objective are expected next year.

FUNDING
NIFA

PROJECT CONTACTS
Greg Hunt (ghunt@purdue.edu)
Marla Spivak (spiva001@umn.edu)
Thomas Webster (thomas.webster@kysu.edu)
Katherine Aronstein (Kate.Aronstein@ars.usda.gov)
Christina Grozinger (cmgrozinger@psu.edu)

Goal 5: Improve the Regulatory Framework to Better Protect Against the Introduction of New Pathogens, Pests, and Parasites of Bees to Meet World Trade Organization (WTO) and International Committee of the World Organization for Animal Health (OIE) Requirements for the Importation and Exportation of Honey Bees

1. Develop new molecular detection systems that can be used to detect pathogens, pests, and parasites in introduced bee stocks and bee products used in beekeeping.

2. Explore opportunities to change regulations based on new molecular detection systems. For example, if a virus is shown to be pathogenic and reliable new detection methods are found, movement of virus-contaminated bee stocks may be regulated by bodies such as OIE.

ACCOMPLISHMENTS

New Molecular Assay for Resistant Bacterial Strains Developed. ARS identified the mechanism of resistance of the bacterial honey bee pathogen (for American foulbrood) to the antibiotics oxytetracycline (OTC) and developed a new molecular assay for detecting OTC-resistant bacterial strains in bee samples. This new identification method is now available for regulatory officials considering ways of preventing the spread of resistant bacterial strains in bee shipments.

FUNDING
ARS

PROJECT CONTACT
Katherine Aronstein (Kate.Aronstein@ars.usda.gov)

Guidelines for Bee Imports Developed. ARS began working with the North American Plant Protection Organization to coordinate regulation and importations of non-Apis bees from Mexico, Canada, and the United States. Initial guidelines were developed for the safe importation of bees to avoid the accidental release of bee pathogens and parasites and to avoid the introduction of bee species that may cause a loss of native bees.

FUNDING
ARS

PROJECT CONTACTS
Rosalind James (Rosalind.james@ars.usda.gov)
James Strange (James.Strange@ars.usda.gov)

3. Establish processes for periodic monitoring of the U.S. honey bee population to determine whether specific pests are present.

ACCOMPLISHMENT

Pilot Survey for Bee Diseases, Parasites, and Pests. APHIS, in collaboration with ARS, conducted a pilot survey in 2009 in Florida, California, and Hawaii to validate sampling protocols used to determine which diseases /parasites/pests of honey bees are and are not present in the United States. A limited national survey is being conducted in 2010 in 13 states.

FUNDING
APHIS
ARS

PROJECT CONTACTS
Robyn Rose (Robyn.I.Rose@aphis.usda.gov)
Jeff Pettis (Jeff.Pettis@ars.usda.gov)

Goal 6: Demonstrate Improved Colony Health by Integrating Research-Derived Knowledge and Tactics into an Areawide Project

1. Test and verify management approaches for mite control, improved diet, improved bee stock, and changes in migratory practice.

ACCOMPLISHMENTS

Areawide Project Initiated. ARS is continuing an Areawide Project on Honey Bee Health across multiple ARS locations (Tucson, Arizona; Beltsville, Maryland; Weslaco, Texas; and Baton Rouge, Louisiana). The project has several aspects, including documenting the impact of migration on bee colonies, examining the effects of supplemental feeding on colony health, developing more resistant bee lines and improved methods for their management, and developing better control methods for honey bee pests. Ultimately, the Project aims to develop a set of best management practices for migratory beekeepers to reduce stress on their bee colonies, thereby enabling bees to ward off threats. Specific results of these individual research projects are described elsewhere in this report.

FUNDING
ARS

PROJECT CONTACTS
Jeff Pettis (Jeff.Pettis@ars.usda.gov)
Tom Rinderer (Tom.Rinderer@ars.usda.gov)
Frank Eischen (Feischen@weslaco.ars.usda.gov)
Gloria DeGrandi-Hoffman (Gloria.Hoffman@ars.usda.gov)

Brood Pheromone Shown to Improve Colony Strength. ARS scientists in Weslaco, Texas, are testing a new brood pheromone device in honey bee colonies to improve the health of honey bees, as well as to improve crop pollination. In a cooperative agreement with Contech, Inc. (formerly PheroTech International Inc.), ARS carried out field trials showing that treated colonies experienced increases in pollen collection and population.

FUNDING
ARS
Contech, Inc.

PROJECT CONTACT
Katherine Aronstein (Kate.Aronstein@ars.usda.gov)

2. Transfer technology for early spring bee availability for pollination.

ACCOMPLISHMENT

Effects of Supplemental Protein Feeding on Colony Growth Assessed. Cranberry pollination is extremely stressful to honey bees, providing minimal nutrition for the amount of work performed. ARS scientists studied whether supplemental feeding with the MegaBee® protein supplement during cranberry pollination could reduce colony losses and improve colony population growth. Results showed that colonies fed the protein supplement grew more than those that were not fed, indicating that even a relatively small addition of supplemental protein to colonies during cranberry pollination improves their growth and survival.

FUNDING
ARS
NIFA

PROJECT CONTACTS
Gloria DeGrandi-Hoffman (Gloria.Hoffman@ars.usda.gov)
Gordon Wardell (GordyAW@aol.com)

Effects of Protein Supplements on Worker Physiology and Immune Response Examined. To evaluate the potential benefits of nutritional supplements on bee health, ARS scientists compared protein levels, endocrine development, and immune response in adult worker bees that were fed protein, pollen, and high-fructose corn syrup supplements. Results showed that protein and pollen supplements produced similarly positive effects, but bees fed high-fructose corn syrup had significantly reduced immune responses. This study will help beekeepers improve management strategies for their bees.

FUNDING
ARS

PROJECT CONTACTS
Gloria DeGrandi-Hoffman (Gloria.Hoffman@ars.usda.gov)
Judy Chen (Judy.Chen@ars.usda.gov)

Goal 7: Transmit or Disseminate Science-Based Information to Manage Bees

1. Develop, maintain, and preserve a secure Web-based site for scientific collaboration (SharePoint).

2. Develop, maintain, and update a Web-based public Internet site, e.g. eXtension or PIPE (Pest Information Platform for Extension).

ONGOING WORK

Best Management Practices Being Established for Bee Conservation. NIFA and ARS (Areawide Project researchers) have facilitated several extension and outreach efforts to inform the public of CCD research findings and updates. In July 2009, NIFA's Coordinated Agricultural Project (CAP) and ARS launched two important new sites: The Bee Health Community of Practice CoP: http://www.extension.org/bee%20health. This is a university eXtension network designed to serve as a clearinghouse for objective, research-based information. The site provides a platform for communities of experts to share articles and resources for download. A Best Management Practices Guide for honey bee health is in progress. For more information on the eXtension site, see Goal 7.

In addition, the University of Georgia developed a new site defining the Managed Pollinator CAP Project including the following: background about pollinator decline and goals, objectives and management plan for this team approach of 22 investigators to reverse pollinator decline. See the CAP project website at www.beeccdcap.uga.edu.

FUNDING
NIFA
ARS

PROJECT CONTACTS
Keith Delaplane (Ksd@uga.edu)
John A. Skinner (Jskinner@utk.edu)
Zachary Huang (Huang@msu.edu)
Greg Hunt (ghunt@purdue.edu)
May Berenbaum (Maybe@uiuc.edu)

ONGOING WORK

Social Media Outreach Initiated. Researchers are pursuing additional outreach opportunities using social networking sites and public Web pages. Honey bee health is featured on YouTube at http://www.youtube.com /beehealth.

Researchers also are now implementing pages on Facebook, Twitter, and Second Life; on beekeeper Web forums such as beesource.com; and on Wikipedia.

FUNDING
NIFA

PROJECT CONTACTS
John Skinner (jskinner@utk.edu)
Keith Delaplane (ksd@uga.edu)

End Notes

[1] The first Pellston Conference was held in 1977 to address the needs and means for assessing the hazards of chemicals to aquatic life. Since then, many conferences have been held to evaluate current and prospective environmental issues. Each has focused on a relevant environmental topic, and the proceedings of each have been published as a peer-reviewed or informal report. These documents have been widely distributed and are valued by environmental scientists, engineers, regulators, and managers because of their technical basis and their comprehensive, state-of-the-science reviews. The first four Pellston conferences were initiated before the Society of Environmental Toxicology and Chemistry (SETAC) was effectively functioning. Beginning with the 1982 conference, however, SETAC has been the primary organizer, and SETAC members (on a volunteer basis) have been instrumental in planning, conducting, and disseminating conference results. Taken from http://www.setac.org/node/104.

[2] Neonicotinoids are a class of pesticide compounds registered on a wide variety of commercial agricultural crops.

[3] van Englesdorp et al, 2010.

INDEX

A

B

D

E

N

O

P

Q

R

S

T

U

V

W

Y